Bite-Size
Science

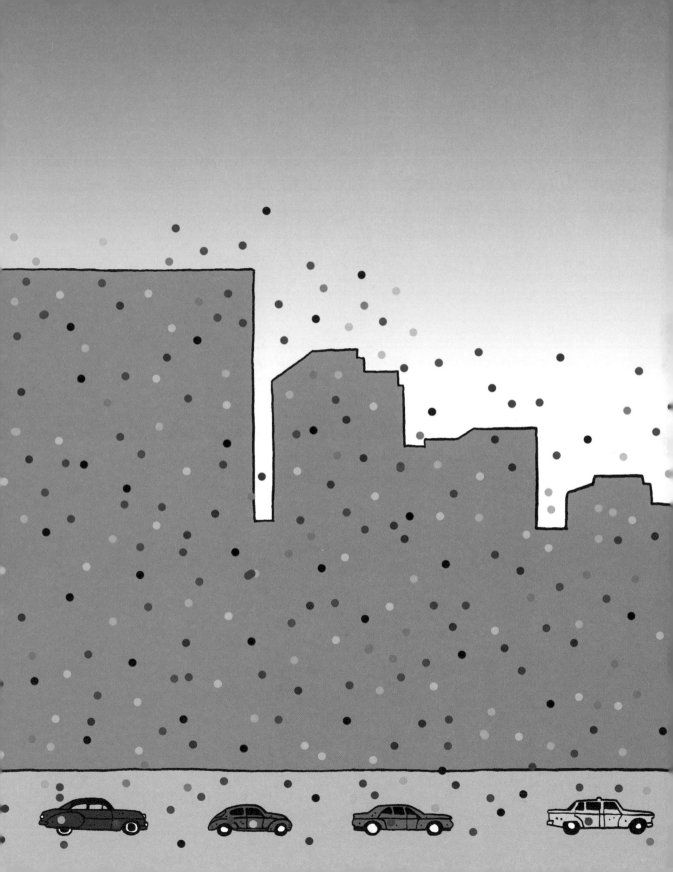

Bite-Size Science

**Everything You Need to Know About Science in
Small, Easily-Digestible Portions**

Robert Dinwiddie

BARRON'S

A QUARTO BOOK

First edition for North America published in 2010 by
Barron's Educational Series, Inc.

All inquiries should be addressed to:
Barron's Educational Series, Inc.
250 Wireless Boulevard
Hauppauge, NY 11788
www.barronseduc.com

ISBN-10: 0-7641-4422-7
ISBN-13: 978-0-7641-4422-6
Library of Congress Control Number:
2009938888

QUAR.WWL

Conceived, designed, and produced by
Quarto Publishing plc
The Old Brewery
6 Blundell Street
London N7 9BH

Project editor: Chloe Todd Fordham
Designer: James Lawrence
Concepts: Paul Carslake
Illustrator: Michael Chester
Additional text: Steve Parker
Copy editor: Cathy Meeus
Proofreader: Liz Dalby
Design assistant: Saffron Stocker
Art director: Caroline Guest
Creative director: Moira Clinch
Publisher: Paul Carslake

Color separation by Modern Age Repro House Ltd,
Hong Kong
Printed in China by 1010 Printing International Ltd

10 9 8 7 6 5 4 3 2 1

Contents

Introduction

Bite-Size Science has been written principally for people whose scientific knowledge is a little hazy or rusty but who would like to "get up to speed" with the types of science they encounter in newspapers, on TV and the Internet, at work, or in conversations with friends. This kind of science might be anything from vaccine safety to the latest research being carried out in underground particle accelerators (and what the point of that research is), or from new discoveries in genetics to the pros and cons of biofuels versus other forms of renewable energy.

In particular, I hope that this book may improve readers' knowledge about the science relating to topics that provoke a lot of controversy—such as the causes of climate change, the desirability or otherwise of stem-cell research, how nuclear waste is to be disposed of, or animal testing of medicines—so that they can more easily form opinions based on knowledge, and engage in arguments around these issues with confidence. Wherever possible, the facts and arguments on either side of these controversial issues have been summarized in an objective way, with the aim of allowing readers to develop their own opinions and come to their own conclusions.

Scope of the Book

By necessity, *Bite-Size Science* is selective. No attempt has been made to cover the whole of science comprehensively, as this would have needed a volume many times larger. Instead the emphasis has been on covering a relatively restricted number of topics in a "bite-size" way, providing enough information to give a flavor of what each topic is about, but without so much detail as to cause indigestion.

I have chosen subjects for inclusion based on a number of different factors including, among others:

• Being in the news, talked about a lot, or controversial (for example: global warming, flu pandemics, genetically-modified foods, DNA fingerprinting, drugs and sport, stem-cell research, and the weather).

• Having a high intrinsic interest because of the unusual, bizarre, or extreme nature of the subject matter (for example: black holes, dark matter, quantum theory, mass-energy, and cloning).

• Sheer scientific importance. Many topics have been included simply because they are vital to the understanding of vast areas of science and so could hardly be left out—although these topics all happen to be of fundamental interest anyway. This applies, for example, to the coverage of atoms and elements (basic to the whole of chemistry and much of physics), plate tectonics (the most important idea in Earth science), Darwinian evolution (a major unifying theme in biology), and the Big Bang (a fundamental idea in both physics and astronomy).

• Coinciding with my personal interests, obsessions, and knowledge base (or more accurately, the gaps in my ignorance).

Structure of the Book

The subject matter has been split into nine chapters. Of these, the first two cover some basic science relating to matter and energy—the two main types of "stuff" in our world. These are followed by chapters on space, origins (how the world came to be as it is, starting from the Big Bang), planet Earth and the life it supports, the environment, human health, and genetics. A final chapter is devoted to some notoriously tricky-to-understand but inherently fascinating topics in modern physics, including relativity and "theories of everything."

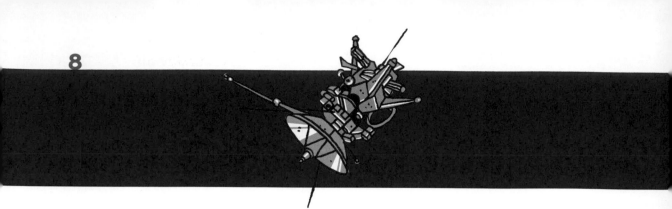

Jargon-Busters and Illustrations

The emphasis throughout this book is on clarity, with the intention of making complex scientific subjects more understandable. Technical terms have been avoided wherever possible, but where unavoidable a "jargon-buster" box has been included nearby with an explanation of the full meaning of the term.

Liberal use has been made of illustrations of various types, including "back of the envelope"-style diagrams, step-by-step sequences, and even cartoons in order to clarify the explanation of the topic under discussion, from the structure of an atom to the origin of the Solar System. Many of these illustrations rely on analogies (several of which happen to be based on food items, for which no apology is made—after all this book is called *Bite-Size Science*!). In the description of Earth's structure, for example, our planet has been represented as a cut-through hard-boiled egg. This might seem odd or even plain silly to the reader at first, but as it happens, the shell, white, and yolk of an egg match to a remarkable degree the proportions between Earth's internal layers—its crust, mantle, and core— and act as a useful reinforcement to this information. Continuing the food theme, the splitting of an atom has been likened to cleaving an apple, and the expansion of the universe to a loaf of raisin bread rising in the oven. Moving away from food, different influences on health have been represented as a card game, plate tectonics has been likened to an airport conveyor system, and so on. To support and reinforce the analogies, more conventional sequenced diagrams have also been used.

Units of Measurement

In this book, most "everyday" units expressing distances, temperatures, weights, and so on have been stated in imperial units (e.g. feet, degrees

Fahrenheit, pounds) rather than in metric units (e.g. meters, degrees Celsius, kilograms). The reason for this is that, whereas metric units are the standard units used by the scientific community, and often for science teaching in schools, the majority of you reading this book are probably not formal students of science. Most likely, you will be more comfortable with temperatures stated in degrees Fahrenheit, and lengths and distances (especially those greater than an inch) stated in feet, yards, or miles.

In a limited number of instances, you will find that metric units have been used rather than imperial units. This includes, for example, very small distances, pertaining to the diameter of atoms or the wavelengths of light waves, where millimeters or nanometers (millionths of millimeters) have been used. This is to avoid expressing the relevant distances as very tiny and hard-to-imagine fractions of an inch.

For readers who prefer to think in metric units rather than imperial units, the brief list of conversion factors shown in the panel on the right may be of use for carrying out conversions between, for example, miles and kilometers, or between degrees Fahrenheit and degrees Celsius.

Robert Dinwiddie
October 2009

Converting Imperial Units to Metric Units

This brief list of conversion factors may help those unfamilar with imperial units.

• An inch is about 2.5 centimeters or 25 millimeters

• A foot is about 0.3 meters or 30 centimeters

• A mile is about 1.6 kilometers

• A pound is about 45 grams or 0.45 kilograms

• A US fluid ounce is about 30 milliliters

• A US pint is about 473 milliliters

• A US gallon is about 3.8 liters

• A 1 degree change or difference in degrees Fahrenheit is about the same as a 0.555 degree change in degrees Celsius. To convert a temperature in degrees Fahrenheit to degrees Celsius, first subtract 32 and then divide by 1.8. To convert a temperature in degrees Celsius to degrees Fahrenheit, first multiply by 1.8 and then add 32.

1

Very Small Stuff

or the Nature of Matter

CHAPTER CONTENTS

This chapter is about matter, one of the two main types of "stuff" that exist in our world, the other being energy. Matter and energy are usually thought of as being distinct entities, although—as we shall see when we come to look at energy in detail (Chapter Two)—they are actually very closely related. One of the main aims of this chapter is to describe what sorts of things matter is composed of, but before delving into that, an important question to ask is—what exactly do we mean by matter?

Defining Matter

A useful way of defining matter is that it is something that possesses both a volume and a mass. Of these concepts, volume is the more straightforward to understand: it is simply the property of taking up space. Mass is slightly trickier. Physicists usually define it in one of two ways. The first defines mass as something that affects the rate at which an object changes its motion when subjected to a force (a push or pull). In other words, mass is the property that affects how difficult it is to shift an object. Alternatively, mass can be defined as the property of an object that

determines how much gravitational attraction it exerts on other objects. In short, matter can be thought of as anything that takes up space, requires effort to shift, and that exerts a gravitational pull on other matter.

What is Matter Composed of?

The best-understood type of matter—though not the only type, or even the most common—is made of the tiny objects called atoms, or their sub-components. The existence of atoms has been suspected for over 2,500 years but it was only in the 19th century that scientists gathered convincing evidence that they definitely exist. By the early 20th century, it was realized that atoms have an internal structure. This chapter starts off by looking at this structure and then considers radioactivity, which is caused by flaws in the structural stability of atoms. Next we look at how atoms combine to make larger structures such as molecules and crystals, then at the elementary particles that make up atoms. The final pages consider dark matter—a mysterious form of matter known to exist in the Universe in large quantities, but whose nature is poorly understood.

• Atoms are made of protons, neutrons, and electrons. Electrons are much smaller than protons and neutrons.

• The central nucleus (containing the protons and neutrons) comprises around 99.9 percent of an atom's mass.

• The width or diameter of atoms varies between about 50 and 500 billionths of a millimeter according to type or element.

• As well as 94 naturally occurring elements, by 2009, atoms of around 23 other elements had been made artificially.

Atoms and Elements

Elements and Compounds

For centuries, chemists have known that certain substances can be decomposed—by, for example, strong heating or passing an electrical current through them—into other substances. These decomposable substances came to be known as compounds. Everyday examples include water, chalk, and common salt. Other substances, such as carbon, copper, and oxygen, didn't seem to be decomposable, and these came to be called elements.

Once it was confirmed that matter consists of atoms, it was also realized that each element consists of a specific type of atom, distinct from the atoms of other elements. Today it is known that all atoms of a particular element have a defining property in common—the same number of protons in their nuclei. The number of protons determines what is called the element's atomic number. Today it is known that there are 94 naturally occurring elements, from hydrogen (atomic number 1) up to plutonium (atomic number 94). Some of these occur only in tiny amounts in nature.

Just about everything on Earth—indeed, in the whole universe that scientists know much about—is made up of atoms, of which there are about 90 different types, each corresponding to a particular chemical element. For example, there are hydrogen atoms, oxygen atoms, carbon atoms, and atoms of gold.

Atomic Structure

Until the late 19th century, atoms were imagined to be small, hard spheres, like miniature pool balls, with no internal structure. But in 1897 a British physicist, J.J. Thomson, showed that atoms can sometimes be made to emit minuscule electrically charged particles that are even smaller than the atoms themselves. The emitted particles soon came to be known as electrons. This discovery immediately raised the possibility that atoms are made up of smaller components, and further experiments over the next 30 years confirmed this to be the case. Atoms are now known to consist of a dense central region called the nucleus, which contains particles called protons and neutrons, and a region surrounding the nucleus called an electron cloud, in which the electrons move.

ELECTRON:
Each electron has about $\frac{1}{2000}$ the mass of a proton or neutron, and is negatively charged. In an isolated, uncharged atom, the number of electrons is the same as the number of protons in the nucleus. Depending on the element, this can be anything from 1 to over 100.

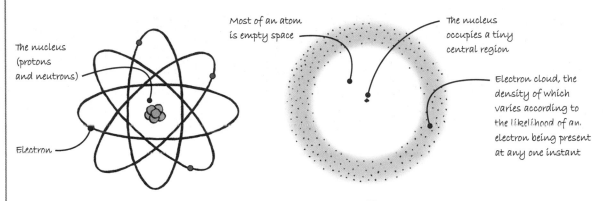

The nucleus (protons and neutrons)

Electron

The Myth

A common old-school textbook depiction of an atom, from the mid-20th century, shows quite a large nucleus at the center, with electrons orbiting the nucleus much in the same way that artificial satellites orbit the Earth today.

Most of an atom is empty space

The nucleus occupies a tiny central region

Electron cloud, the density of which varies according to the likelihood of an electron being present at any one instant

The Reality

In reality, the nucleus is much smaller than usually depicted, and electrons cannot be said to occupy particular locations or orbits—although as they move around, they are more likely to be located in some regions of the atom than others.

Jargon buster

NUCLEUS: The nucleus of an atom consists of positively charged protons and uncharged neutrons. The numbers of these depend on the chemical element of which it is an atom.

The Analogy: A Stadium-Sized Atom

If an atom had a diameter of 500 feet—roughly the dimensions of a large sports stadium—then its nucleus would be about the size of the ball in the referee's whistle. This is true for the atoms of nearly all elements, although for just a handful, the nucleus would be slightly larger or smaller. The electrons would be no larger than specks of dust, moving around anywhere within the stadium, although they would be more likely to be located at some particular distances from its center than others.

What Holds Atoms Together?

Electrons, being negatively charged, are held in atoms by electrostatic attraction toward the nucleus, which is positively charged because of the presence of protons. The nucleus itself is held together by the mighty nuclear force.

How is it that protons, which should not hang out together at close quarters, do so in the nucleus of every atom?

Inside an atomic nucleus, an intense conflict plays out between two powerful forces. These are the electromagnetic force—which causes like charges (such as protons) to repel and unlike charges to attract—and the nuclear force, which pulls protons and neutrons together. Normally, the nuclear force wins out...

THAT DAY IN THE NUCLEUS...

1 I HATE YOU ALL. YOU'RE REPELLENT. GOODBYE!!!

2 THIS ELECTROMAGNETIC FORCE IS DRIVING US APART...

EFFECTS OF THE ELECTROMAGNETIC FORCE
Because the electromagnetic force causes like electrical charges (positive-positive or negative-negative) to repel, the positively charged protons in the nucleus are constantly pushing away from each other.

3 HANG ON...

EFFECTS OF THE NUCLEAR FORCE

Fortunately, the even more powerful nuclear force overrides the electromagnetic force and pulls all the protons and neutrons together. If this didn't happen, the world as we know couldn't exist, because any atoms that formed would immediately break up. The nuclear force is closely linked to (some physicists consider it a "leakage" of) yet another force that operates inside protons and neutrons known as the "strong" force (see Composite Particles, page 23).

4 I CAN'T GO ON! SOMETHING IS DRIVING ME BACK!

5 THE NUCLEAR FORCE WAS TOO MUCH FOR US...

6 BUT I STILL HATE YOU ALL AND FIND YOU REPELLENT!

A LOVE-HATE RELATIONSHIP

Thus, the protons and neutrons are held tightly in a compact ball. Most nuclei remain like this forever, or at least for billions of years. Only in a minority does the conflict between the electromagnetic and nuclear forces eventually cause the nucleus to break up, in what is called radioactive decay (see page 16).

• Radioactivity is the result of instability in the nuclei of atoms of certain isotopes (forms of elements).

• Radioactive isotopes emit energetic particles and sometimes hazardous gamma radiation of varying degrees of intensity.

• One gram of the intensely radioactive isotope radium-226 undergoes about 37 billion nuclear transformations every second.

• Because the instability in a radioactive isotope resides in the nuclei of its atoms, it cannot be made non-radioactive by chemical treatment.

Radioactivity

What are Isotopes?

Each chemical element can exist in a number of different forms called *isotopes*. These vary in having different numbers of neutrons in their atomic nuclei. Isotopes have names like oxygen-14 or lead-206—the number after the name of the element denotes the total number of protons and neutrons in the atomic nuclei of that isotope. Some isotopes are stable, others unstable. For example, carbon-12—by far the most common isotope of carbon—is stable, while carbon-14—a much less common isotope—is unstable and will eventually decay.

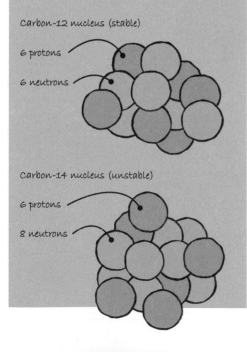

Carbon-12 nucleus (stable)

6 protons

6 neutrons

Carbon-14 nucleus (unstable)

6 protons

8 neutrons

Not all atoms are stable. While most are likely to last for hundreds of billions of years, some contain a particular combination of protons and neutrons in their nuclei that makes them unstable— this is because of conflicts between the forces operating inside the nuclei (see page 14). As a result, the nuclei of these atoms eventually disintegrate, or at least change in some way—in a process called radioactive decay.

Radioactive Decay

Unstable atoms that are liable to undergo radioactive decay are called radioactive isotopes—isotopes being forms of chemical elements (see *What are Isotopes?*, left). Most naturally occurring elements have several unstable radioactive isotopes as well as one or more stable forms. A few with particularly large atomic nuclei, such as uranium, radium, and all artificial elements, exist only as radioactive isotopes.

When they decay, radioactive isotopes emit energetic particles and sometimes also emit some high-energy radiation (see *Types of Radioactive Decay*, right). As they decay, they are themselves transformed into the atoms of a different element, and often generate a lot of heat, and sometimes light as well—a pellet of radium, for example, glows in the dark and produces heat for centuries. Sometimes the product of radioactive decay is itself unstable, leading to a series of unstable isotopes that decay into one another to form "decay chains."

⇨

Types of Radioactive Decay

When radioactive isotopes decay, their nuclei usually emit one of two types of particles, called alpha and beta particles. They may also give off gamma rays. These three types of emission vary in how energetic, penetrating, and hazardous to health they can be.

Alpha particle
(two neutrons and two protons): not very penetrating and not dangerous unless source is eaten or inhaled

Alpha Decay

In alpha decay, emission of an alpha particle means that the atomic nucleus of the daughter isotope has two fewer protons than the parent nucleus. Consequently, the daughter is an isotope of a different element from the parent—one with an atomic number two lower than that of the parent. For example, when an isotope of polonium (atomic number 84) undergoes alpha decay, the daughter is an isotope of lead (atomic number 82).

Beta particle (electron): can penetrate living tissue and may cause damage

Beta Decay

In beta decay, as an electron flies out of the parent nucleus, one of the neutrons (shown here as light blue) changes into a proton (green). Because this increases the number of protons in the nucleus by one, the daughter is an isotope of a different element from the parent—one with an atomic number one higher than that of the parent. Thus, when an isotope of carbon (atomic number 6) undergoes beta decay, the daughter is an isotope of nitrogen (atomic number 7).

Gamma rays:
exposure to any more than small doses of these can be a serious health hazard

Gamma Decay

Gamma radiation is a highly energetic form of electromagnetic radiation (see page 32)—a type of wave rather than a particle. Radioactive isotopes never emit gamma rays on their own but frequently emit them alongside alpha or beta particles. Unlike alpha and beta decay, the emission of gamma radiation has no effect on the number of protons or neutrons left in the nucleus. Instead, gamma ray emission converts the nucleus from a higher, unstable energy state into a lower, more stable one.

- Some radioactive isotopes have half-lives measured in thousands of trillions of years, an example being selenium-82.

- About 80 percent of an average person's exposure to radiation results from naturally occurring radioactivity.

- This average level of radiation exposure is, however, about 100 times less than the level of exposure needed to raise cancer risk by one percent.

- Uses of radioactive isotopes range from detecting and treating diseases to smoke alarms, paper production, and nuclear energy.

Half-Lives

Different radioactive isotopes vary enormously in how unstable they are and therefore how quickly they decay. A standard way of categorizing the stability (or instability) of a particular isotope is by reference to its *half-life*. The atoms of a radioactive isotope don't all decay at once after a set amount of time. Instead, they decay one by one, in a random pattern. The half-life of an isotope is the time that can be expected for half of a collection of its atoms to decay. Very unstable isotopes have extremely short half-lives, measured in fractions of a second, while the more stable ones have half-lives measured in thousands, millions, or even billions of years.

There is a real problem with what to do with material containing long half-life isotopes produced as waste by nuclear reactors. This is because the material remains radioactive—and therefore potentially hazardous—for such a long period of time.

Radioactivity Exposure from Natural Sources

Everyone is exposed to a low level of radioactivity from natural sources, including radioactive substances in the ground and in the air. The main source of such radiation is a colorless radioactive gas called *radon* that seeps out of the ground, where it is formed from the radioactive decay of uranium. Higher levels of radon are leaked from certain rock types, such as granite. In regions where there is a large proportion of these types of rock in the ground, the gas can accumulate to hazardous levels in poorly ventilated buildings. For anyone living in a part of the world where relatively high levels of radon are known to seep from the ground, it is a good idea to have its concentration in any poorly ventilated rooms (such as cellars and basements) professionally checked from time to time.

Radioactivity Exposure from Human Activity

In addition to this exposure to natural radioactivity, everyone is exposed to a little additional radioactivity in the environment that has been caused by human activity over the past 100 years or so—mainly as a result of nuclear bomb tests in the atmosphere, actual use of such bombs in the past, and accidents or leaks that have occurred from nuclear power plants. Little can be done to avoid exposure to this radioactivity. An additional source of exposure occurs in some people from medical procedures (diagnostic or treatment), but in these situations the procedure is never suggested as an option unless the potential benefits outweigh any risk from exposure to such levels of radiation.

Radioactivity—Bad or Good?

Although radioactivity is often thought of as being harmful, in reality radioactive isotopes are used in many different ways in the modern world, and provide many benefits. At the crux of any debate over their use is the extent to which any particular application can be safely controlled to prevent environmental contamination or risks to human health. Most uses are noncontroversial, but one that has attracted passionate debate for decades is energy from nuclear fission (see page 35), which uses radioactive isotopes as a fuel.

Aspect	How and Why?
Health hazards	Exposure to moderate to high levels of radioactivity increases the risk of developing cancer, because radiation can damage the DNA in cells causing them to multiply out of control. Damage to DNA in sperm or egg cells (or their precursors) can result in genetic disorders in offspring.
Medical treatment	The gamma radiation emitted by some radioactive isotopes can be used to treat cancers by killing the cancer cells.
Medical diagnosis	Radioactive isotopes can be injected into the bloodstream and their subsequent progress through the body monitored in order to locate cancers and diagnose other serious medical problems.
Renewable energy source	Earth's natural radioactivity creates heat, which is tapped by geothermal power stations as a non-polluting renewable energy source (see page 112).
Sterilization	Gamma rays from radioactive isotopes can be used to kill microbes on objects such as surgical instruments, and also in food to increase shelf-life.
Industrial uses	Radioactive isotopes are commonly employed in industry—for example, in paper-making to check paper thickness. This is based on paper of different thicknesses absorbing different amounts of emitted particles.
Home uses	Smoke alarms employ a radioactive isotope that emits alpha particles. These help set up a small electric current in the detector. Smoke particles disrupt the transmission of the current, thereby setting off an alarm.
Dating technology	Measuring the radioactivity in items ranging from rocks to old bones and tools can be used to date them, which is a tremendous aid to geological and archeological research.
Other scientific research	Various radioactive isotopes are used for studying molecular structures and chemical processes occurring in animals and plants.
Nuclear energy	Some radioactive isotopes are used as a fuel for nuclear reactors. These reactors have an advantage over fossil-fuel power plants in that they provide energy without releasing much carbon dioxide into the atmosphere (which is the main cause of global warming). Against that, there is: an identifiable risk of hazardous radioactive material being released into the environment; the largely unresolved problem of how to dispose of the radioactive waste (see page 114); and many other issues.

- Molecules are two or more atoms joined together. Crystals are huge numbers of atoms joined together in a repeating structure.

- Atoms can be joined by ionic bonds—formed after electrons have moved between atoms—or by covalent bonds, in which electrons are shared between atoms.

- The number of known types of molecules runs into millions. There is no limit to the number of new ones that could be made.

Crystals and Molecules

Most atoms don't exist on their own, but instead are joined to other atoms. The "ties" or attachments between these atoms are called *chemical bonds*. There are two basic types of chemical bonds. *Ionic bonds* occur in large, repeating structures called crystals. *Covalent bonds* firmly link two or more atoms into a structure called a *molecule*—a configuration of joined atoms with a distinct shape, but not so large that it is classed as a crystal. A few molecules contain atoms of only one element, but a far greater proportion contain atoms of two or more elements. Most compounds (substances composed of more than one element) exist either in the form of crystals or molecules. Whatever form it takes, a compound usually has properties that are completely different from those of its constituent elements. Sodium chloride (better known as common salt) is nothing like either sodium, an unstable gray metal, or chlorine, a pungent-smelling green gas.

Chemical Formulas

Every compound has a formula—a sort of shorthand that identifies what types of atoms (i.e. elements) the compound contains, and in what ratio. For example, the formula of methane is CH_4. This indicates that it contains carbon (C) and hydrogen (H), with four hydrogen atoms for each carbon.

Ionic Bonds

For ionic bonds to form between atoms, first something special has to happen—the atoms have to become electrically charged, either by gaining or by losing electrons. When this happens, they become charged particles called *ions*—an atom that loses one or more electrons becomes a positively charged ion, and an atom that gains one or more electrons becomes a negatively charged ion. Once ions have formed, the process of *electrostatic attraction* binds them together to form a crystal. Ionic bonds are simply the electrostatic forces of attraction between the ions in a crystal.

1. DONATING AND ACCEPTING ELECTRONS

Atoms of sodium have a tendency to lose an electron through interactions with other atoms, because this makes their internal structure more stable. Atoms of chlorine similarly have a tendency to gain an electron. So when atoms of these elements come into contact, it is almost inevitable that an exchange of electrons will occur. And indeed, an exchange does occur.

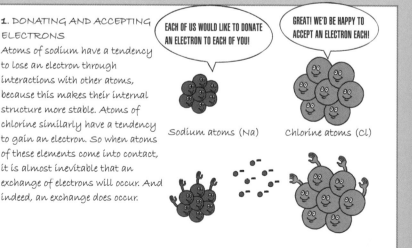

EACH OF US WOULD LIKE TO DONATE AN ELECTRON TO EACH OF YOU!

GREAT! WE'D BE HAPPY TO ACCEPT AN ELECTRON EACH!

Sodium atoms (Na)

Chlorine atoms (Cl)

Covalent Bonds

In covalent bonds, instead of electrons moving between atoms, the electrons are shared. In general, atoms will readily share electrons if the molecule formed is more stable, and has an overall lower energy level than the uncombined atoms. This means that when the atoms combine, some energy is released as heat. Atoms can also sometimes be made to join where their energy (once they have combined) is greater than that of the uncombined atoms. For this to happen, energy (in the form of heat) has to be added to trigger the joining process.

For most pairs of elements there is a "preferred" or most stable configuration. For hydrogen and oxygen, for example, the preferred, most stable combination is H_2O (water). Another less stable configuration is H_2O_2 (hydrogen peroxide). Other configurations, such as H_3O, have no reason to exist in nature, since they would be much less stable than water and would tend to quickly decompose into it.

1. A MEETING OF THREE ATOMS

An oxygen atom can achieve greater stability by sharing some of its electrons with other atoms, as long as those other atoms can also provide electrons for sharing. In this case, each hydrogen atom supplies one electron for sharing, and the oxygen atom supplies two electrons for sharing—one to share with each hydrogen atom. The oxygen atom is said to have formed a single covalent bond with each atom of hydrogen.

2. MOLECULE FORMATION

The three atoms have joined together to form a molecule of water or H_2O—two hydrogen atoms joined to one atom of oxygen. The electron clouds of the two hydrogen atoms have partly merged with the outer of two electron clouds surrounding the oxygen nucleus.

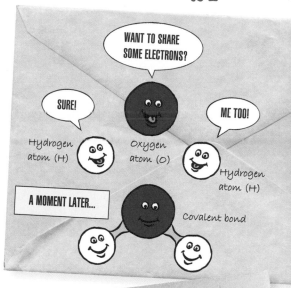

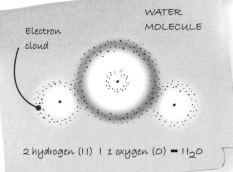

WATER MOLECULE

2 hydrogen (H) + 1 oxygen (O) = H_2O

---

2. ATTRACTION

Once electrons have been exchanged, each sodium ion carries a positive charge—it is now a positive ion. Conversely, each chlorine ion has become a negative ion. These ions are now attracted to each other by the electromagnetic force.

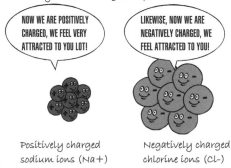

Positively charged sodium ions (Na+)

Negatively charged chlorine ions (Cl-)

3. CRYSTAL FORMATION

In an instant, the sodium and chlorine ions arrange themselves into a repeating three-dimensional structure; they have formed a crystalline compound called common salt or sodium chloride (NaCl). A crystal of salt, much more extensive than shown here, can contain a vast number of ions.

A MOMENT LATER...

Crystal of sodium chloride (common salt)

• Approximately 60 different elementary particles have now been identified, as well as scores of composite particles (consisting of joined elementary particles).

• The *Higgs boson* is a yet-to-be confirmed particle that is thought to play a part in giving other particles their mass.

• Among the known elementary particles are *neutrinos*, trillions of which are estimated to pass through the human body every second.

Particle Physics

Particle physics is the study of particles that are smaller than atoms. We've already looked at three types of particle in this category: protons, neutrons, and electrons. Until about 80 years ago, these were thought to be the only subatomic particles, and few scientists imagined that there might be any others. But since then, hundreds of new ones have been discovered, many of them extremely unstable.

ANTIPARTICLES AND ANTIMATTER: Most subatomic particles have equivalent antiparticles, which are identical except that they possess an opposite electrical charge. For example, the antiparticle of the electron is the positron, which is the same as an electron but positively charged. Collectively, all the different types of antiparticle are called antimatter. When particles and antiparticles meet, they annihilate each other to produce pure energy. For reasons that are still being investigated, there appears to be far less antimatter than matter in the Universe.

Elementary Particles

Quarks

Up quark

Down quark

Charm quark

Strange quark

Top quark

Bottom quark

Other Particles

Electron

Neutrino

Gluon

Photon

The "Zoo" of Particles

The first few new particles were discovered from studying what happens when cosmic rays—highly energetic particles from space—collide with matter on Earth. Later, more were revealed when known particles were smashed together in particle accelerators. By around 1970 the new particles that had been discovered were collectively being referred to as a particle "zoo."

To make sense of all these new particles, as well as their interactions, a new theory in physics called the Standard Model was devised. According to this model, subatomic particles can be divided into *elementary particles*, which have no substructure, and *composite particles* (or *hadrons*), which are made of elementary particles joined together.

The most important elementary particles that make up matter came to be known as *quarks*, and these are now known to exist in six different types (known as quark "flavors"). At the same time, it became clear that protons and neutrons are actually composite particles made up of triplets of quarks. Other elementary particles that form part of the Standard Model include electrons, other low-mass particles called *neutrinos*, and "messenger" particles, known as gauge bosons. These play roles in various interactions between matter particles and have other influences on the underlying nature of matter. But there is one particle that forms part of the Standard Model that has yet to be observed: the Higgs boson (see opposite page).

Composite Particles

Also called *hadrons*, these are composed of quarks or of quarks and their antiparticles (antiquarks) (see Jargon buster, opposite). The two shown here—the familiar proton and neutron—are now known each to be composed of three quarks held together by what is called the strong force. This force is mediated by "messenger" particles called *gluons*.

Up quark

Down quark

Gluon

A proton contains 1 down and 2 up quarks, held together by gluons.

A neutron contains 1 up and 2 down quarks, held together by gluons.

Large Hadron Collider

The Large Hadron Collider (LHC) at CERN (the European Organization for Nuclear Research) in Switzerland is the world's biggest experimental apparatus for research into particle physics. It is a huge particle accelerator, a machine that accelerates beams of subatomic particles (in this case, the class of particles called hadrons, which includes protons) to velocities close to the speed of light and then smashes them together. The object of this is to recreate the types of high-energy conditions that previously occurred only in the very early stages of the development of the Universe, just after the Big Bang. Physicists are confident that by creating these conditions, it should be possible to discover more about the fundamental nature of matter and the interactions that occur between particles.

The Higgs Boson

One of the objectives of the LHC is to demonstrate—or alternatively to rule out—the existence of a particle called the Higgs boson, which (if it exists) gives other particles their mass. Many physicists think that soon after the Big Bang, an invisible force field called the Higgs field was formed, together with the Higgs boson. Any particles that interacted with this field acquired a mass via the Higgs boson. The more they interacted, the more mass they acquired, whereas particles that never interacted with the field were left with no mass.

Because the future direction of particle physics hinges on whether this explanation of the origin of mass is correct, answering this question has become an urgent matter for physicists, and is the main reason why so many resources have been poured into developing the LHC. In addition, research carried out using the LHC may help to answer other unresolved questions: whether the different forces that govern nature (such as the forces that hold atoms together) are part of a single unified force that existed in the first moments after the Big Bang; what the nature of dark matter might be (see page 24); and whether the Universe possesses more spatial dimensions than we currently know about.

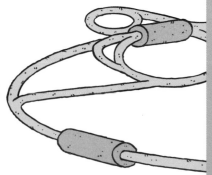

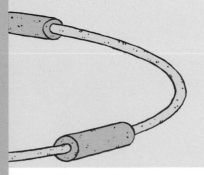

LHC Fact File

• Circumference of tunnel: 44 miles

• Depth: 165–500 feet underground

• Number of large magnets required to keep particle beams aligned: 1,600

• Estimated total cost of project: $4.6–9 billion

• Number of times per second a proton will travel around the tunnel during operation: 11,000

- About 22 percent of the total mass in the Universe is believed to consist of mysterious "dark" matter. This is more than the mass of the visible matter, made of atoms.

- The remainder of the mass of the Universe is believed to exist in the form of a mysterious phenomenon called dark energy, which seems to be causing the Universe to expand at an ever-increasing rate.

- Dark matter might consist of dead stars, dust, black holes, particles called neutrinos, or other, unknown forms of matter.

Dark Matter

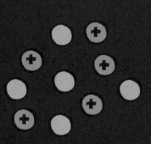

Although modern science knows an awful lot about atoms, and much about particles smaller than atoms, oddly, it doesn't know what most of the matter in the Universe is made of. For over 70 years astronomers have been aware that, in addition to the visible material observable in the Universe, there is a large amount of some mysterious invisible material, which has been labeled "dark matter."

What Might Dark Matter Consist of?

There are a number of candidates for what dark matter might be. If it consists mainly of particles that are currently unknown to science, then these may be revealed by experiments using machines such as the Large Hadron Collider.

COLD ORDINARY MATTER
Dark matter might consist of the type of matter we are familiar with, but which is invisible because it gives off little light or radiation. Dead or faint stars (called brown dwarfs), planets, or clouds of cold dust in space could consist of such dark matter.

NEUTRINOS
Neutrinos have a very small mass and are difficult to detect, but scientists are sure that there are very large numbers of them swarming around in the universe, and their combined mass might be sufficient to account for some of the dark matter.

BLACK HOLES
The majority of black holes, which by their very nature are difficult to detect, are remnants of stars that exploded in the past. It is possible that enough of them exist as remnants of the earliest stars to account for some of the dark matter.

WIMPs
Dark matter might consist of particles that science is unable to detect in any way at present. These particles have a name—WIMPs (weakly interacting massive particles)—even though nothing is known about them except that they possess mass.

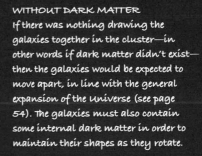

Evidence for Dark Matter

Most of the evidence for the existence of dark matter comes from the study of galaxies. Galaxies slowly rotate, and in order to maintain themselves intact without some of the material at their edges spinning off into space, they must contain enough mass to produce sufficient internal gravitational attraction. However, the visible stars in galaxies don't possess enough mass to generate such force, so scientists have concluded that there must be something else supplying the "missing" mass. A similar argument applies to clusters of galaxies; these stick together despite an apparent lack of mass to prevent them from drifting apart. Again, dark matter filling the voids between the galaxies provides the explanation.

WITHOUT DARK MATTER
If there was nothing drawing the galaxies together in the cluster—in other words if dark matter didn't exist—then the galaxies would be expected to move apart, in line with the general expansion of the universe (see page 54). The galaxies must also contain some internal dark matter in order to maintain their shapes as they rotate.

WITH DARK MATTER
Something invisible is holding galaxy clusters together by gravitational attraction. It can only be something that possesses mass, since only material that has mass exerts a gravitational pull. Yet telescopes and other devices have failed to find anything in the voids between galaxies that gives off any type of radiation.

Gravitational attraction

Dark matter

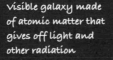

visible galaxy made of atomic matter that gives off light and other radiation

2

Energy Stuff

CHAPTER CONTENTS

The Universe is filled with energy. This chapter explores the main forms of energy, including light and other electromagnetic radiation as well as different types of stored or "potential" energy. It also looks at the special link between matter and energy that physicists have termed "mass-energy."

What is Energy?

Unlike matter, energy cannot be placed on a laboratory bench for study and analysis, but we all have a rough idea of what it is: a quality that makes some objects or people more "active" or "dynamic" than others. Scientists define energy as the ability to do work. In less formal terms, energy is something that allows "stuff to be done." Thus, energy moves our cars along the highway, heats and lights our homes, powers our computers, and plays sounds on our digital music players. Energy from the Sun heats Earth's oceans and atmosphere, causing phenomena such as winds, ocean currents, clouds, and rain. Sunlight is also the ultimate source of energy in the food we eat, which in turn provides the energy that allows us to move, breathe, grow, and think.

Energy Transformations

Over the course of history, people have learned how to convert energy from one form into another for many useful purposes. The discovery of fire turned the energy held in wood and (later) coal into heat and light; with the birth of the Industrial Revolution, new machines for transforming energy were developed; and at the advent of the Nuclear Revolution, scientists discovered that part of the mass of an atomic nucleus was convertible into vast amounts of energy.

In all energy transformations, no energy is gained or lost overall, but the amount of useful energy produced (whether from a bicycle, a cannon, or a steam engine) is always less than the amount of energy input, because some energy is lost—in the form of heat—to the environment. The proportion of useful energy obtained is referred to as "energy efficiency." Scientists are striving to develop new technologies with superior energy-efficiency. Continuing advances in this area may well make a significant contribution to how the world will meet its future energy needs.

- There are two main categories of energy: energy based in movement and potential (or stored) energy.

- Energy can neither be created nor destroyed, but only converted from one form into another.

- Scientists define energy as the ability to do work and they measure it in units called *joules*.

- Everyone's personal (bodily) energy expenditure depends on their weight and what daily exercise they perform.

Forms of Energy

WORK AND ENERGY: Energy is what allows work to be done, and work converts energy from one form into another. Work is performed when a force (with a strength measured in units called *newtons*) is applied over a particular distance (measured in meters).

JOULES: Both work and energy are measured in units called *joules*. A joule is the amount of work done when a force of 1 newton is applied over 1 meter. For example, to push a heavy crate across a wooden floor, the force required might be 800 newtons. If it is pushed for two meters, the work done would be 800 x 2 = 1600 joules.

KILOCALORIE: The kilocalorie is another unit of energy, commonly used to express energy values in food. A kilocalorie (kcal) is equal to about 4.2 kilojoules (kJ), or 4,200 joules. A Calorie, with a big "c," is the same as a kilocalorie.

POWER: Power is the rate of work, or the rate at which energy is transformed from one form into another. Its unit is the watt, which equals an energy conversion rate of one joule per second. A bulb with a power rating of 100 watts is converting electrical energy into light and heat at the rate of 100 joules per second.

Energy comes in many different forms, which fall into two main categories. Movement-based forms of energy are those possessed by objects because of their own motion or because of the motion of particles inside the object. Potential energy includes stored energy and the energy an object possesses simply because of its position in relation to other objects.

Movement-Based Energy

The simplest form of energy based on movement is called *kinetic* energy, and is the energy possessed by an object travelling from one place to another, such as the energy in a moving ball or in the wind. *Thermal* energy (heat) is manifested by the movement and vibrations of atoms and molecules. *Electrical current* energy is the energy of electrons moving through metal wires or other conductors. *Ocean-wave* energy is the energy in waves moving across the surface of the sea, while *sound* energy consists of waves of compression and rarefaction (decompression), which can move through any substance. *Radiant* energy, including visible light and other types of energy in the form of electromagnetic waves (see page 32), can also be considered a movement-based form of energy.

Potential Energy

One of the more obvious types of *potential* energy is elastic energy—energy stored in an object through the application of force, such as the energy in a compressed spring or a stretched rubber band. *Chemical* energy is the energy stored in the bonds between atoms in substances such as petroleum, or in fats and sugars in the human body. *Gravitational potential* energy is the

HOW IS ENERGY CONVERTED?
A mass of rock possesses gravitational potential energy, which dissipates when it falls under the force of gravity.

As the rocks fall with increasing speed toward the truck, their gravitational potential energy decreases, but at the same time another form of energy—the kinetic (movement) energy of the material—progressively increases. One form of energy (gravitational potential) is being converted into another form of energy (kinetic).

energy possessed by an object purely by virtue of its position relative to other objects—for example, a rock at the top of a hill has more gravitational potential energy than the same rock at the bottom of the hill. Another type of potential energy is stored in the nuclei of atoms, and provides the basis for nuclear power (see pages 34–35).

Energy Conversions

Energy can neither be created or destroyed—a principle known as the *law of conservation of energy*. Whenever work is done, energy is neither lost nor gained, but instead is converted from one form into another. For example, if a person pushes a heavy object along the ground, energy is converted from a form stored chemically in the person's muscles into heat energy. This heat energy appears as a slight warming of the ground and the base of the pushed object. In the example given on this page of rock tipping, gravitational potential energy is converted first to kinetic energy (the energy of movement) then to heat and sound energy as the rocks hit the truck. A kettle transforms electrical energy into heat, while a TV converts it into light, sound, and heat. Over the whole planet, energy is constantly being converted from one form into another on a vast scale. Plants constantly convert radiant energy from the Sun into stored chemical energy. Solar radiation falling on the atmosphere and oceans causes warm, moist air to rise and form clouds. Later, when the water of which the clouds are composed falls as rain, its potential energy as it flows down over the land may be captured by a hydroelectric power station and converted to electrical energy.

When the rocks hit the back of the truck, another energy conversion occurs. The material loses its kinetic energy as it stops moving, but this energy reappears in two other form. A little is released as sound energy—there's a fair amount of noise. But most reappears as heat. If the temperature of the rocks were to be measured, they would be found to be a few degrees warmer than they were at the top of the cliff.

Your Personal Energy Expenditure

Everyone has a personal daily energy expenditure, which depends mainly on what types of activity he or she engages in, and for how long each day. It also depends on his or her weight. If the expenditure of energy is greater than the energy the person takes in each day in the form of food, his or her weight will decrease. If the reverse is true, his or her weight will increase.

Running or Similar High-Intensity Exercise

Jogging very briskly or engaging in other high-intensity exercise, such as high-energy sports, expends 10–25 kilocalories per minute, depending on your weight.

To calculate your energy expenditure in kilocalories each day from high-intensity exercise, divide your weight in pounds by 11, and then multiply that figure by the number of minutes each day you engage in that exercise.

Cycling or Other Moderate-Intensity Exercise

Cycling at a moderate speed, or other moderate intensity exercise, such as moderate swimming, expends 5–12 kilocalories per minute, depending on your weight.

To calculate your energy expenditure each day from moderate-intensity exercise, divide your weight in pounds by 22, and then multiply that figure by the number of minutes each day you engage in that exercise.

From Sunlight to Lightbulbs

Radiant energy from the Sun...

...is converted by the process of photosynthesis within the leaves of plants into energy-rich chemicals.

The plant dies, is buried, and over millions of years turns into coal, which retains the chemical energy that was originally stored by the plant.

Walking or Other Low-intensity Exercise

Walking at a moderate to low speed, or other low-intensity exercise, such as gardening, golf, or housework, expends 160–440 kilocalories per hour, depending on your weight.

To calculate your energy expenditure each day from low-intensity exercise, multiply your weight in pounds by 1.8, and then multiply that figure by the number of hours each day you engage in the low-intensity exercise.

Sedentary Activities

When not exercising, such as when sitting watching TV, at a computer, or reading, you expend 60–165 kilocalories per hour, depending on your weight.

To find your energy expenditure each day from sedentary activities, divide your weight in pounds by 1.5, and then multiply that figure by the number of hours each day you spend on sedentary activities (excluding sleeping).

Figure It Out

To get a rough idea of your energy expenditure per day, calculate how many hours (or partial hours) you engage in high-intensity, moderate-intensity, low-intensity, and sedentary activities per day. Taking into account your weight, use the preceding guidelines to calculate your daily expenditure on these activities, and add them up. Then add on the amount of energy you expend while asleep, as shown at right, because even when sleeping your body uses some energy just keeping itself alive. This gives you your total energy expenditure.

Energy Used When Sleeping

Your Weight in lbs (kg)	Kilocalories
90 (40)	450
110 (50)	480
130 (60)	520
155 (70)	560
175 (80)	600
200 (90)	640
220 (100)	680
240 (110)	720

In a power plant, the chemical energy in the coal is converted into thermal energy possessed by heated water, then into the kinetic energy of steam.

A turbine converts the kinetic energy of steam into electrical energy.

In the home, electrical energy is converted into heat, light, and sound by appliances such as kettles, light bulbs, and televisions.

• Electromagnetic (EM) radiation includes a number of related forms of energy including light, radio waves, microwaves, and X-rays.

• All EM radiation travels through empty space at the same speed—the speed of light—which is about 300 million meters per second.

• EM radiation travels as waves which differ in their wavelength (size) —from less than a billionth of a millimeter to thousands of miles— according to type.

• As well as having the properties of waves, EM radiation also behaves in some ways like a stream of energetic particles.

Electromagnetic Radiation

RADIO WAVES
Radio waves are the form of radiant energy with the longest wavelengths. They divide into several wavelength bands, including long-wave, medium-wave, short-wave, and VHF (very high frequency; also very short wavelength). For medium-wave and long-wave radio broadcasting, the radio signal is carried by varying the amplitude of the broadcast waves; this is called amplitude modulation or AM. For VHF the signal is transmitted by varying the frequency of the waveform. This is called frequency modulation or FM.

One of the more ubiquitous and useful forms of energy is radiant energy, which is also called electromagnetic (EM) radiation. This comes in a range, or spectrum, of different forms, which includes radio waves, microwaves, visible light, infrared and ultraviolet radiation, X-rays, and gamma rays. The energy we receive on Earth from the Sun is all radiant energy, mainly visible light with some infrared and ultraviolet. What all these energy forms have in common is that they travel as a series of waves through space and some matter. The basic difference between them is in the distance between the waves (called the wavelength of the radiation) as they travel along. All forms of EM radiation travel at the same high speed—the speed of light— in a vacuum.

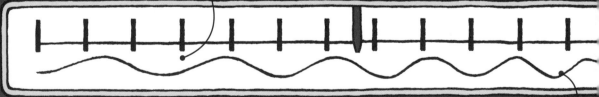

Volume

There are two different aspects to the strength or "volume" of EM radiation. First, the different types—radio waves, visible light, X-rays, and so on—each have a different intrinsic intensity, or energy level, which increases with decreasing wavelength. Thus, gamma rays and X-rays are intrinsically more energetic than, for example, light or radio waves. Second is the sheer amount of the radiation being poured out by the energy source—the difference in visible light, for example, between a flashlight and a searchlight.

MICROWAVES
After radio waves, the EM waves with the next longest wavelengths are microwaves. These are used for a variety of technologies including mobile phones, microwave ovens, some TV broadcasts, Bluetooth and Wireless Lan ("Wi-Fi") communications, and GPS (Global Positioning System) devices.

Light Waves and Color

Visible light occupies a small part of the electromagnetic spectrum—the part comprising wavelengths between 400 and 700 nm (nanometers or billionths of a meter). Different regions within this overall band of wavelengths correspond to different colors—for example, red light has a wavelength of around 680–700 nm while blue light has a wavelength of between 420 and 450 nm. Only pure colors of the rainbow—red, orange, yellow, green, blue, and violet—are associated with particular wavelengths. All other colors, such as shades of brown, pastels, and so on, arise from mixtures of wavelengths. This includes white, which is a roughly equal mixture of all the rainbow colors.

Wavelength

Electromagnetic radiation consists of fluctuations in the strength of electric and magnetic fields. These fluctuations move in a particular direction. In a particular type of radiation, such as blue light, all the peaks and troughs in the train of waves are regularly spaced. The distance between adjacent peaks is called the wavelength of that radiation.

VISIBLE LIGHT
This small part of the EM spectrum is sandwiched between infrared and ultraviolet waves.

ULTRAVIOLET WAVES
Used in devices such as sunbeds and some lasers, ultraviolet is relatively short-wavelength, high-energy radiation.

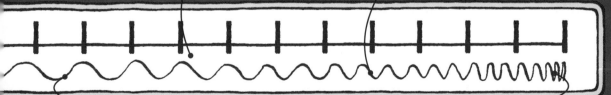

INFRARED WAVES
Next up from microwaves are infrared waves, which include radiant heat but also find uses in devices such as TV remote controls.

X-RAYS AND GAMMA RAYS
X-rays, as used in hospitals, and gamma rays, produced by some radioactive substances, are the shortest-wavelength, highest-energy types of EM radiation.

Radiant

• Mass and energy can be thought of as different aspects of the same quantity, called mass-energy.

• A tiny amount of mass always converts into an enormous amount of energy.

• The conversion of mass to energy occurs in both nuclear fission and nuclear fusion.

• Nuclear fission is the process that powers nuclear power plants, while nuclear fusion powers the Sun.

Mass-Energy and Splitting the Atom

Probably the most remarkable form of energy of any sort is nuclear energy. As generated in nuclear power plants, this relies on *nuclear fission*—a process in which the atomic nuclei of elements such as uranium are caused to split up in a chain reaction, causing a release of energy. Nuclear energy is often referred to as the "energy that resides in atoms," but more accurately it should be described as "the energy that comes from mass."

An illustration of what is meant by this is shown on the opposite page, using the analogy of an apple being chopped in two with a cleaver. When the apple is chopped, a tiny bit of apple comes up "missing," but is later found to be stuck to the cleaver. When an atom is split, a little bit of mass really does disappear—it turns into energy! This is explained by a principle first proposed by Albert Einstein in 1905, called mass-energy equivalence, which states that mass and energy are not separate entities, but two different aspects of a single quantity called mass-energy. The fact that mass and energy are equivalent means that mass can turn into energy, and energy into mass. Einstein devised his famous formula (below) to show the relationship between mass and energy in terms of the speed of light in a vacuum. As this speed is so large (over 186,000 miles per second, or nearly 300 million meters per second), it follows that a tiny bit of mass is equivalent to a colossal amount of energy, and this is why processes such as nuclear fission generate such huge amounts of energy.

Why Specify a Vacuum?

References to the speed of light usually mean its speed in a vacuum (empty space). Light travels faster through a vacuum than through anything else. When traveling through matter, it slows down a little, to about 200 million meters per second—when passing through glass, for example.

$$E = mc^2$$

Einstein showed that mass and energy are connected by this famous formula where E is energy, m is mass and c is the speed of light in a vacuum.

Suppose the 2 grams of apple that ended up on the cleaver (opposite) were turned completely into energy.

The amount of this energy would be:
0.002 (mass in kilograms)
x (300 million)^2
(speed of light squared)

= 180 trillion joules. That is enough energy to power more than 5,000 American homes for a year!

"THE ENERGY THAT COMES FROM MASS"
Imagine that the nucleus of a uranium atom is actually an apple, with a mass of 150 grams (about 5.3 oz).

Then picture the apple being split into two parts by chopping it with a cleaver—the equivalent of the process of fission (to split a uranium nucleus a neutron is fired into the nucleus; this destabilizes the nucleus so that it splits).

The apple is now in two parts. A few bits have been released (in a uranium atom a few neutrons are released, which continue the chain reaction). But when the two parts of the apple and the bits are collected together, their combined mass comes to only 148 grams: two grams of the apple's original mass seem to have disappeared!

It is discovered that the missing mass is a tiny two-gram fragment of apple that has become stuck to the edge of the cleaver. But with the splitting of a uranium nucleus, the missing mass has turned into energy.

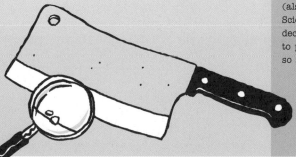

From Mass to Energy

There are two different ways in which changes within the nuclei of atoms can lead to the conversion of small amounts of mass to a large amount of energy. These are:

- the splitting of nuclei into smaller fragments: nuclear fission;

- the combining of nuclei to make larger nuclei: nuclear fusion.

Fission

Nuclear fission occurs naturally in radioactive decay (see pages 16–20). It is also artificially-induced as an energy-generating process in nuclear power plants. This artificially induced fission involves setting up controlled chain reactions in a suitable (fissile) material (usually an isotope of uranium). Fission is also the source of explosive power in fission bombs (formerly known as atom bombs).

Fusion

Nuclear fusion occurs naturally in the Sun and other stars. In fact, fusion processes that occur in the Sun, which produce solar energy, are the ultimate source of most energy resources on Earth. Because fusion in stars builds up large atomic nuclei from smaller nuclei, over billions of years, it has also been a major source of the atoms that make up our world. Only the atoms of the very lightest chemical elements, hydrogen and helium, were not originally forged by fusion in stars (see pages 62–63).

On Earth, fusion is the source of the explosive power in hydrogen bombs (also called H-bombs or fusion bombs). Scientists have been working for decades on a way of harnessing fusion to provide controlled release of energy, so far with only limited success.

3

Very Big Stuff
or the Observable Universe

CHAPTER CONTENTS

This chapter looks at topics related to the world outside our planet: space or the visible Universe, as it appears from Earth today. The first few pages provide a reminder of the vast distances and numbers involved in describing the Universe, as we move outward from Earth and its moon to the other planets and our local star, the Sun, to other nearby stars, then out to the whole of our galaxy (the Milky Way), and finally beyond it to other galaxies.

The Workings of the Universe

We then look at what goes on in the Universe. Many topics are covered, including the structure of the Milky Way, the lives and workings of stars, and cycles in the activity of the Sun that are believed to have effects on Earth. The possible results of star death are considered, including supernovae, neutron stars, and the bizarre objects known as black holes. We touch on the nature of gravity and the mechanics of orbits, the logistics of sending probes to explore the Solar System, and the evidence that the whole Universe is gradually expanding. The final pages briefly consider the likelihood that life exists beyond Earth.

What is the "Observable" Universe?

The observable Universe is simply what we observe by looking into space with our naked eyes or using telescopes or other instruments. For a number of reasons, the observable Universe is not exactly the same thing as the entire present-day Universe. The most obvious reason for this is that most matter in the Universe is invisible—the so-called dark matter described in Chapter One. Another is that, because of the vast distances involved and because light has a finite speed, we are unable to observe most objects in space as they are right now. Instead we see them as they once were. When we observe stars, we see most of them as they were tens, hundreds, or thousands of years ago, and distant galaxies as they were millions or billions of years ago. It is possible that there are vast regions of the Universe that have been rushing away from us so fast since the Universe began that light from them has never reached us—they are beyond a "horizon of visibility"— and in this case the whole Universe is bigger, far bigger, than the "mere" observable Universe.

• Our "local neighborhood" in space is the Solar System—the Sun and everything orbiting it. The Sun accounts for 99.8 percent of the mass of this neighborhood.

• The average distance from Earth to our very nearest neighbor, the Moon, is about 30 times Earth's diameter. It takes a spacecraft three days to make the journey.

• The part of the Solar System containing the planets is nearly six billion miles across and would take more than a decade to traverse in a spaceship.

• Beyond the Solar System is an expanse of nothing, until you reach the next nearest star after the Sun; more than 25 trillion miles away.

Our Backyard in Space

Space is truly vast. We live in a tiny corner of it—in our Solar System, which in turn is part of the Milky Way, a galaxy containing billions of stars. And beyond the Milky Way, there are tens of billions of other galaxies. The dimensions and distances of even the parts we can see are so vast that they are literally beyond our understanding. For example, the Sun is 865,000 miles in diameter and is 93 million miles from Earth.

How can we understand these numbers? Suppose the 93 million miles between the Sun and Earth is re-sized as a 75-meter running track. The Sun is on the starting line and is about 27 inches in diameter—about the size of a large rubber ball. Planet Earth, which is a little over ¼ inch in diameter (the size of a pea), is at the finish line. The moon is less than ¹⁄₁₂ of an inch across (the size of a small grain of rice). On our running-track scale, the Moon is in orbit about eight inches from Earth. In the 1960s and 1970s, Apollo spacecraft took three days to cover that distance.

The AU

The distance between Earth and the Sun, shown below as the 75-meter running track, is also known as one astronomical unit (AU). It's a handy way of expressing big distances. One thousand AU is 1,000 times 93 million miles—in other words, 93 billion miles.

The distance from Earth to the Sun is shown as a 75-meter running track. On this scale, the Sun (on the starting line) is about 27 inches in diameter.

This string takes you to the next nearest star. If the distance from the Earth to the Sun was the width of these two open pages...

...then this string would need to be about 60 miles long!

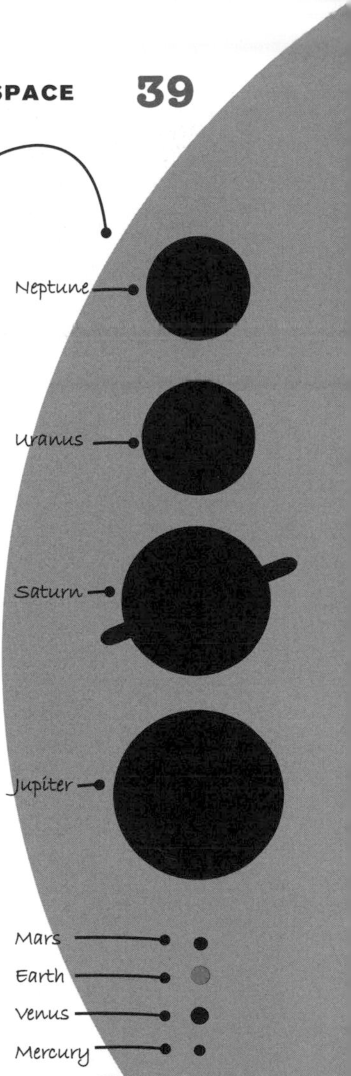

On our scale, we can now position other planets in the Solar System. Mercury is a very small pea about 95 feet from the starting line. Venus is about the same size as Earth and about 70 feet from the finishing line (it took five months for the Venus Express space probe to make the journey from Earth to Venus). Mars is off the end of the track, 130 feet beyond Earth. Jupiter is the size of a tennis ball, out in the parking lot, and Saturn is nearly the same size, a few blocks from the stadium (it took a little over three years for the space probe Voyager I to reach it). Uranus is the size of a walnut, about a mile from the stadium, and Neptune is a similar size, about one and a half miles away.

How Far to the Nearest Star?

Our Sun is just one of several hundred billion stars in the Milky Way. Yet even the nearest of all these other stars, Proxima Centauri, is impossibly distant. The nearest star would be 13,000 miles from our "stadium." If the distance from the Earth to the Sun was the width of these two open pages, then to connect the Sun to the nearest star you would need a a piece of string nearly 60 miles long!

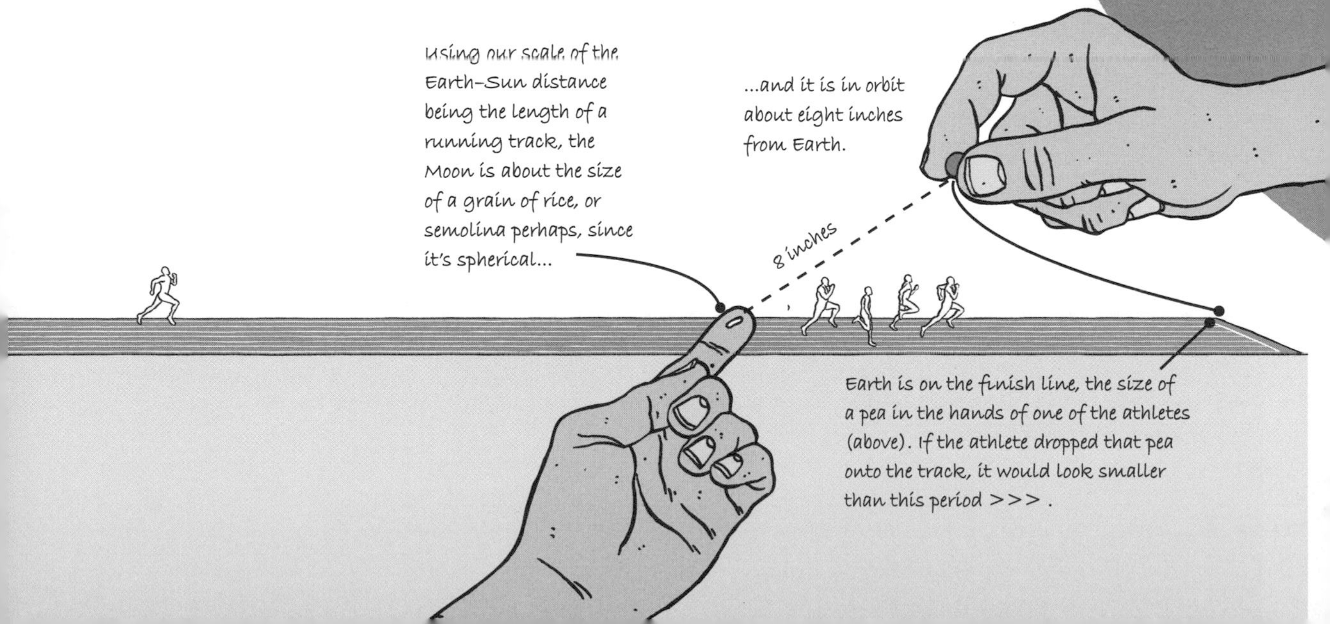

RELATIVE SIZES
Measured by diameter, the Sun (orange) is about 110 times bigger than Earth. Comparing the Sun's size, by diameter, to that of the other planets, it dwarfs them by factors ranging from 10 times for Jupiter, the largest planet, to over 280 times for Mercury, the smallest.

Neptune

Uranus

Saturn

Jupiter

Mars
Earth
Venus
Mercury

Using our scale of the Earth-Sun distance being the length of a running track, the Moon is about the size of a grain of rice, or semolina perhaps, since it's spherical...

...and it is in orbit about eight inches from Earth.

8 inches

Earth is on the finish line, the size of a pea in the hands of one of the athletes (above). If the athlete dropped that pea onto the track, it would look smaller than this period >>> .

bite size facts

- Our galaxy is just one of an estimated 80 billion galaxies in the visible Universe.

- The light from some of the faintest, most distant galaxies has taken 13 billion years—most of the age of the Universe—to reach us.

- Spiral galaxies, like our own, are one of three main galaxy types: others are elliptical (shaped like an egg) or irregular in shape.

- Globular clusters, found in the outer regions or "halo" of our galaxy, are fuzzy-looking spherical objects, each holding up to a million stars.

Our Galaxy and Beyond

Jargon buster

LIGHT-YEAR: the distance that light travels through space in one year: about 5.9 trillion miles (5,900,000,000,000).

FROM YOUR HOME TO DEEPEST SPACE IN 14 STAGES

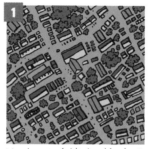

Your home neighborhood is about a mile across and includes your home and local amenities.

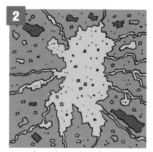

A typical city might be about six miles across.

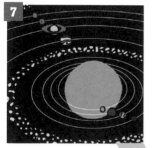

The outer Solar System (six billion miles across) includes the orbits of all the planets.

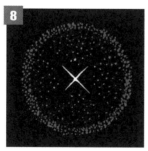

The Oort cometary cloud that surrounds the Solar System is about two light-years across.

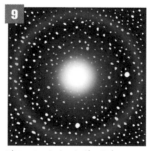

The nearest star to the Solar System is Proxima Centauri, 4.35 light-years away.

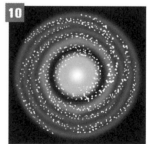

100,000 light-years across, the Milky Way has a dense central bulge and several spiral arms.

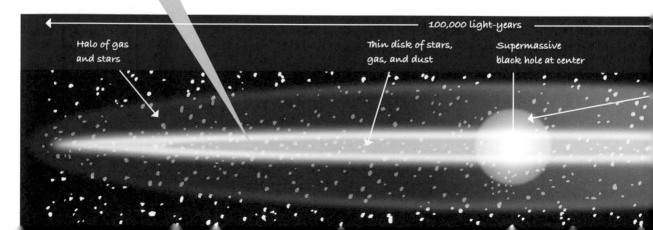

◄———————————— 100,000 light-years ————————————

Halo of gas and stars

Thin disk of stars, gas, and dust

Supermassive black hole at center

The Universe has a hierarchy of structures. The Earth is in the Solar System, which is part of our galaxy, the Milky Way, which in turn is just one of a local group of about 50 separate galaxies. The local group of galaxies contains the Milky Way, along with one other large spiral galaxy (the Andromeda galaxy), and numerous smaller galaxies. It is just part of a collection of galaxy groups called the local supercluster. The observable Universe—the spherical region from which light has had time to reach the Earth since the beginning of the Universe—is about 90 billion light-years in diameter. Beyond this is (possibly) a larger region from which no light has ever reached us, and perhaps never will.

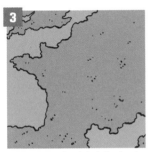

An average-sized state is about 300–500 miles across.

Earth is a rocky sphere about 8,000 miles in diameter. It has surface water and supports life.

The diameter of the Earth-Moon system is 450,000–500,000 miles (the Moon's orbit is not circular).

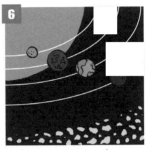

The inner Solar System (60 million miles across) includes the Sun and its four inner planets.

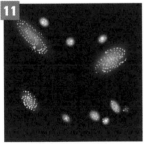

The local group of galaxies, which includes the Milky Way, is about 10 million light-years across.

The local supercluster, called the Virgo Supercluster, is 100 million light-years across.

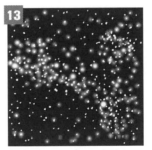

The local group of superclusters is one billion light-years across. It contains the Virgo Supercluster.

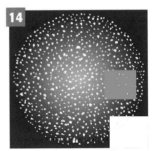

The visible Universe consists of galaxy clumps and voids tens of billions of light-years across.

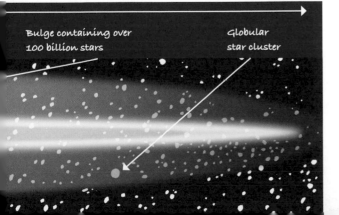

Bulge containing over 100 billion stars

Globular star cluster

The Milky Way—Our Galaxy

Viewed from the side, the Milky Way can be seen to consist of a central bulge and a surrounding thin disk. The bulge contains over 100 billion stars, and is known to have a supermassive black hole at its center. Surrounding the bulge and the disk is a "halo" of stars and also many large, spherical clusters of stars (globular star clusters).

• Our galaxy contains about 200–400 billion stars, but only about 9,000 of these are visible individually to the naked human eye—even to a person with perfect vision under ideal conditions.

• All stars that we can see individually (as well as huge numbers of fainter stars) are situated within about 4,000 light-years of the Solar System.

• The stars in any constellation are all situated roughly in the same direction as seen from the Earth, but aren't necessarily closely grouped together in three-dimensional space.

Constellations

On a dark night, away from any light pollution, thousands of stars can be seen from Earth. All the brighter ones (and many not-so-bright ones) have been grouped into constellations—patterns of stars that (using some imagination) can seem to resemble animals, such as lions or scorpions, objects, such as cups and arrows, or mythological figures, such as Hercules. But, although the stars in a constellation may look close together as seen from Earth, in three-dimensional space they are often located hundreds of light-years apart.

NORMA-CYGNUS ARM
This arm has two parts: the Norma region is the part nearest the center of the galaxy (much of it cannot be seen from Earth); the Cygnus region is the outermost part.

PERSEUS ARM
A new space telescope, the Spitzer telescope, has enabled us to see that this spiral arm and the Scutum-Crux arm are more prominent than the other arms and contain a higher density of stars.

100,000 light years

YOU ARE HERE
This region within the Orion arm, between the yellow arrows, has the Solar System at its center and contains all the stars that are visible individually to the naked eye. It is situated about 25,000 light-years from the center of the galaxy.

The Visible Stars

The most obvious objects in our galaxy—because they produce so much light and other radiant energy—are stars. These huge balls of plasma (electrically charged gas) generate their energy by nuclear reactions and shine as a result of this massive production of energy. Nearly all the stars we can see individually in the night sky are situated within just one small part of one arm of the Milky Way galaxy. Several of these spiral-shaped arms—four major ones and a few minor ones—stream outward from the bulge of stars at the galactic center.

SCUTUM-CRUX ARM
Because it wraps around the back of the center of the galaxy, much of this major spiral arm cannot be seen from Earth. The Scutum region is the part nearest the galactic center; the Crux region is the outermost part.

SAGITTARIUS ARM
So-named because a large part of it lies in the same direction as the constellation of Sagittarius, this arm contains some nebulae (gas clouds) and clusters of stars that are visible from Earth through a good pair of binoculars.

ORION ARM
One of the minor arms of the Milky Way, the Orion arm branches off from the nearby Sagittarius arm. About 30,000 light-years long and 1,000 light-years deep and wide, it contains some dust and large clouds of hydrogen gas as well as billions of stars.

Star Life-Cycles

Stars are continually forming in our galaxy—and other galaxies—from large clouds of gas that contract gradually under the influence of their own gravity. As one of these clouds contracts, its center becomes denser and hotter, until eventually it becomes so hot that nuclear fusion reactions (see page 35) start up. These produce vast amounts of light, heat (which keeps the nuclear reaction going), and other forms of electromagnetic radiation. This radiation pours out in the form of tiny packets of energy called photons, which create a type of pressure pushing outward from the center of the star. This maintains the star's size and shape by preventing its gravitational collapse.

Just about everything about a star, such as its initial size and color, its lifespan, and its eventual fate, depends on its starting mass. The longest-lived stars have a very low mass. These make up the majority of stars and stay as dull, small red stars for tens of billions of years. Average-sized yellow stars (for example, our Sun), have shorter lifespans, while the very largest and hottest stars—huge white- and blue-colored giants—have much shorter lives and die spectacularly (see pages 46–47).

- The Sun produces a constant stream of charged particles, the solar wind, which flows through space at over 600,000 miles per hour.

- Regular variations in activity at the Sun's surface are known as the solar cycle. This cycle can have effects on the Earth's climate.

- Sunspots are relatively cool, dark areas on the Sun's surface. The number of these, depending on the stage in the solar cycle, can be from 0 to over 100.

- The temperature of the Sun varies from 27 million °F in its core to around 10,500°F at its visible surface.

The Sun and its Cycles

Our local star, the Sun, is a fairly average yellow star: a place of extreme temperatures and energy output. At its center, or core, which has a temperature of 27 million °F, nuclear fusion produces a vast amount of energy. This is carried outward by radiation and convection, and then released at the surface as heat and light. In addition, the Sun emits vast amounts of other types of radiation, including ultraviolet, X-rays, gamma rays, and radio waves, and also a constant stream of electrically charged particles called the solar wind.

Solar Cycles

The interior of the Sun is affected by a strong magnetic field that results from the fact that the Sun is a rotating body composed largely of charged particles. Different parts of the interior rotate at different speeds, causing the magnetic field lines to become twisted and entangled over time. This entanglement produces various types of disturbances at the Sun's surface, including sunspots, solar flares, and prominences. These disturbances ebb and flow over cycles that last an average of about 11 years and have some effects on Earth.

Sunspots

The cycles of solar activity are most obvious as variations in the numbers of sunspots—relatively cool, dark areas—on the Sun's surface. Sunspots result from magnetic field lines becoming tangled near the surface and inhibiting heat flow from the interior. Over each cycle, the number of spots varies from zero—during what is called a solar minimum—to as many as 100 during a solar maximum. The positioning of sunspots varies, too; at the start of each cycle they occur well away from the Sun's equator, but as a solar maximum approaches, they become concentrated close to the equator.

FACT: Unusually powerful sunspot activity in October 1991 created such a large disturbance in Earth's magnetic field that it disrupted electrical systems across North America.

A few spots well away from the Sun's equator

Increasing numbers of spots, getting nearer the equator

No spots

Sunspots Over Time

1975　1976　1977　1978　1979　1983　1986　1987　1989　1990

Solar Flares and Mass Ejections

Although sunspots are the most obvious indicators of changes occurring over a solar cycle, other phenomena occurring around the same time are more important for their effects on Earth. These include solar flares and coronal mass ejections. These events blast quantities of charged particles into space, creating the cosmic equivalent of a tsunami.

Effects of Solar Cycles on Earth

During a solar maximum, as occurred in 2001 and is predicted to occur again in 2011–2012, there is a marked increase in the quantity of both charged particles and radiation of all sorts coming from the Sun. When clumps of charged particles hit Earth's magnetic field, they distort it and cause large electrical disturbances in the upper atmosphere. This process causes particularly strong auroras (also known as the Northern Lights in the northern hemisphere). It can disrupt radio transmissions and cause power blackouts and damage to electrical transmission lines and satellites. There is also some evidence that solar cycles may affect weather and climate. A 70-year cold snap in the 17th century, which saw the River Thames in southern England regularly freeze over, coincided with a prolonged drop in sunspot activity. Today, there is some evidence that peaks in solar activity may be linked to heavy rainfall in Africa. However, to date nobody has come up with a convincing explanation for how changes in solar activity can produce such effects. Similarly, although it is often claimed that variations in solar activity affect human mental and physical health, there is as yet no credible theory as to how this might happen.

CORONAL MASS EJECTION: a huge clump of plasma that has broken away from the Sun's atmosphere and is heading off into space.

PHOTOSPHERE: the Sun's visible surface, with a temperature of 10,500°F (this compares with a temperature of 3,000°F for molten steel).

CORONA: a layer of electrically charged gas above the Sun's visible surface, with a temperature of 3,600,000°F.

SOLAR FLARE: a very hot, bright, violent eruption in the Sun's atmosphere caused by twisted magnetic field lines.

SUNSPOTS: relatively cool, dark spots on the Sun's photosphere.

PROMINENCE: a huge loop or cloud of dense, hot gas that hangs above the Sun's visible surface.

The Sun at solar maximum: as many as 100 spots present, mainly near the equator

Diminishing spots

1991 ➡ 1995 ➡ 1997 ➡ 1998 ➡ 1999 ➡ 2000 ➡ 2001 ➡ 2004

- When it runs out of fuel at the end of its life, a low-mass star simply contracts and gradually fades away.

- A medium-sized star, such as our Sun, swells to form a giant red star, then disintegrates to leave a remnant called a white dwarf.

- A high-mass star eventually explodes in an extremely violent event called a supernova and leaves remnants—either a neutron star or a black hole.

- A neutron star is so dense that a pinhead-sized piece of it, if brought to Earth, would weigh more than a blue whale.

When Stars Die

Stars do not last forever, although some continue to shine for tens or even hundreds of billions of years. Eventually, they run out of the hydrogen fuel they need to continue generating energy from nuclear reactions in their core. What happens next depends on the star's mass. Some simply fade away, while others die in spectacular explosions.

becoming a huge star called a red giant. It later collapses again as it switches to burning helium in its core. Eventually, a dying medium-mass star expands again and sheds its outer layers, while its inner regions collapse to form a tiny star called a white dwarf. Eventually, it is thought that white dwarfs will fade into dim objects called black dwarfs.

Low- and Medium-Mass Stars

When a star with a very low mass—much less than that of our Sun—runs out of hydrogen, it simply contracts, cools, and fades away. A medium-mass star, such as the Sun, expands as it burns hydrogen in its outer layers,

High-Mass Stars

The highest mass stars, with more than about six times the mass of the Sun, have very different lives and fates. These stars are especially hot, large, and bright, and burn their fuel at a fast rate, so they have comparatively

Supernova Danger?

Range Limit of a Dangerous Supernova

The heat and radiation from a supernova spreads across a galaxy, but the intensity of the radiation lessens with distance, and in order to have seriously damaging effects on Earth (and life on Earth), it is estimated that a supernova would have to explode within a distance of about 100 light-years—more than 24 times the distance to the nearest star, Proxima Centauri. Fortunately, no star capable of "going supernova" is currently within this region, so we can sleep easily at night!

100 Light-years: Danger Zone!

Earth would escape the dangerous effects of a supernova occurring more than 100 light-years from the Solar System.

Distance from Solar System ⟶   0   100 light-years   200 light-years

short lives. They develop into huge stars called supergiants and go through a series of expansions and contractions during which they burn a succession of chemical elements as fuel. Eventually they disintegrate in fantastically violent explosions called supernovas, which emit more energy than many billions of hydrogen bombs. Supernovas are an important source of some chemical elements, a few of which can only be synthesized in the extreme temperature conditions that occur during these explosions (see page 63).

Sun Life-Cycle

The Sun is a low-to medium-sized star that does not have enough mass to explode as a supernova at the end of its life. It has already existed, and been burning its hydrogen as nuclear fuel, for 4.5 billion years. In about five billion years time, it will start to run out of this fuel and will expand into a red giant. As it does so it will overwhelm Earth and the other planets of the inner Solar System. Within about another billion years, having entirely run out of fuel, it will become a white dwarf, which subsequently will fade over many billions of years to a black dwarf.

Remnants of Supernova Explosions

When a star explodes as a supernova, one of two types of remnant remains:
• **A black hole** (see pages 48–49).
• **Neutron stars** are made entirely of neutrons and have an exceedingly high, but finite, density. One type of neutron star is a pulsar, a fast-spinning object just a few miles across that emits powerful beams of radiation into space. Pulsars oriented in certain directions relative to Earth can be detected by pulsed flashes of radiation that come from them, in a "lighthouse-like" effect (because the pulsar is spinning). In some cases, the regularity of these pulses is more precise than an atomic clock. Pulsars were first detected in the 1960s, but remarkably their existence was predicted long before that, in the 1920s.

Red Giant

White Dwarf

Black Dwarf

The Sun today

E

F

Closest Supernova Candidate

The red supergiant star Betelgeuse is the closest star with a realistic chance of exploding as a supernova in the next 1,000 years. In fact, there is a small chance that it has already exploded and that the light and other radiation hasn't reached us yet. If it did explode, it would outshine the Moon for many months, but it is too far away from Earth (about 600 light-years distant) to have damaging effects on our planet.

Betelgeuse supernova

300 light-years   400 light-years   500 light-years   600 light-years   700 light-years

- A black hole is a region of space that exerts such a strong gravitational pull that no matter or light can escape from it.

- In our galaxy there may be millions of black holes that have formed from the collapse of large stars.

- Astronomers are convinced there is a black hole, with a mass 4 million times greater than that of the Sun, in the middle of our galaxy, the Milky Way.

- Although light cannot escape a black hole, black holes may emit a form of weak radiation called Hawking radiation.

Black Holes

A black hole is a region of space that contains at its center some matter squeezed into a point of infinitesimal size and infinite density. This concentration of matter is called a singularity. In a spherical region around the singularity, the inward gravitational pull is so great that nothing, not even light, can escape—that's why a black hole looks black.

Origins of Black Holes

There are two main types of black hole. A stellar black hole forms from the death of a very large star, which occurs when the star runs out of fuel for its energy-producing processes and explodes in a supernova (see pages 46–47). During this explosion, part of the star implodes inward and just keeps on collapsing until it concentrates matter in a single point. Supermassive black holes are found at the centers of galaxies. These may be a by-product of the process of galaxy formation.

A well in space and time
A black hole can be considered to be like an infinitely deep well—a gravitational well—in space and time. Objects, light, or other radiation that get too close to it may fall in. Here, both matter and light are falling into the hole, never to return.

Near to and inside the black hole, the surface (and thus space and time) is severely distorted because of the extreme concentration of mass at its center.

Detection of Black Holes

Because no matter or light comes out of black holes, they are difficult to detect by normal means. Viewed from a "safe" distance—which would maybe be no closer than a few million miles even if one were traveling fairly fast—a black hole would look like a tiny, black circular disk, very difficult to see against the black background of space. However, a hole should also be detectable from the behavior of matter (such as gas, dust, and stars) close to the hole, as this matter would be gravitationally attracted toward it. So far, astronomers have found several objects in our galaxy that they think are black holes, because gas from nearby stars is being attracted toward these areas and then whirling around and into them. These objects are all thought to be black holes formed from collapsed stars. In addition, astronomers are now fairly certain that there is a supermassive black hole at the center of our galaxy. The evidence for this is based on the behavior of one star, which is orbiting something that is compact and has a very high mass, but that also seems to be invisible.

Surrounding a black hole is a sort of boundary, called the event horizon (shown here in red), which is the point of no return for any matter or light that enters or falls into it.

Properties of Black Holes

Black holes are such unusual objects that physicists have to use an advanced mathematically-based theory, called general relativity (see page 166) to model and describe their properties. Originally developed by Albert Einstein, general relativity proposes that when studying objects like black holes, space and time cannot be considered as separate entities but are linked together in a four-dimensional arrangement called spacetime. General relativity also proposes that the gravity exerted by extremely dense, massive objects like black holes works by distorting both space and time. According to this theory, a black hole can be regarded as an infinitely deep well in the fabric of space and time. Around this well, there is a boundary, called the event horizon. Any object approaching the well too closely and crossing the event horizon, will fall in, be crushed and become part of the singularity at the center. Light and other radiation can also fall into the well (or hole) never to return.

Light or matter falling into the hole spirals down toward its center. As it does so, any matter is torn apart and crushed.

In this representation, the distance down represents increasing strength of gravity, reaching an infinitely high strength of gravity at the center of the well (which is why it is infinitely deep).

bite size facts

• Orbits result from the gravitational attraction between objects interacting with their existing motion.

• Closed orbits in nature are never in the form of a perfect circle, but always have the shape of an ellipse (a stretched circle).

• Because its orbit around the Sun is an ellipse, not a circle, Earth is three percent nearer the Sun in January than in July.

• Some orbits—like the orbits of some comets around the Sun—are open-ended and non-returning, following curves called hyperbolas.

Orbits and Gravity

Many objects in space are in orbit around other objects. These include planets around the Sun, moons around planets, artificial satellites or the space shuttle around Earth, stars around other stars, and the whole Solar System around the center of our galaxy. A classic example of an orbit is that of the Moon around the Earth, the study of which (by Sir Isaac Newton in the mid-17th century) was extremely important in the development of our understanding of the force of gravity and the laws of motion.

Newton realized that the force that causes objects to fall to the ground on Earth might also extend into space and be responsible for holding the Moon in its orbit. By analyzing the motion of objects in space, Newton formulated his law of universal gravitation. This proposed that every object in the Universe exerts a force of attraction—gravity—on every other object and described how the strength of the force between any two objects varies with their masses and the distance between them.

How an Orbit Works

1. Imagine a magical island surrounded by a sea that has some unusual properties.

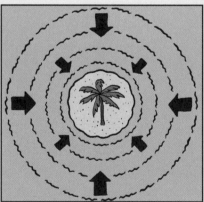

2. All the currents flow toward the island. In addition, the sea exerts no frictional drag on boats passing across its surface, just as space exerts no frictional drag on a spacecraft moving through it.

3. A man in a small boat is passing the island and soon realizes that there is no frictional resistance to the boat's passage. The man turns off the motor to conserve fuel and notices that the rudder is stuck in the "straight ahead" position. This is of no concern as the expectation is that the boat will continue moving in a straight line past the island at its existing speed.

Gravity, Motion, and Orbits

Newton's description of gravity dovetailed with other important laws he formulated concerning the motion of objects. The first of these laws stated that all objects either remain stationary or move in a straight line at a constant speed, unless acted on by a force, such as gravity. The second law stated that when gravity (or any other force) acts on an object, it causes an "acceleration" in that object's motion, which in this context can mean a change in its speed, direction, or both.

When applied to an object such as the Moon in its orbit of the Earth, Newton's laws work as follows: If gravity were suddenly not to exist, the Moon would move in a straight line at a constant speed, which would soon take it a long way from Earth. But because gravity does exist, the Moon experiences a constant gravitational acceleration, or pull, toward the center of the Earth. This deflects what would otherwise be the Moon's straight-line motion into a curved trajectory. Both the direction of the motion, and the direction of acceleration, constantly change, resulting in a near-circular orbital path. Thus, although the Moon is constantly being accelerated toward the Earth, it never reaches it—and will never do so, unless something were to slow it down and change the balance of forces.

Shapes of Orbits

The main difference between orbits is in their shape. All closed orbits (ones that continue around and around) have the shapes of ellipses (stretched circles). Many orbits, such as the Moon's, are close to being circular. But others, such as the dwarf planet Pluto's orbit around the Sun, are very elongated ellipses.

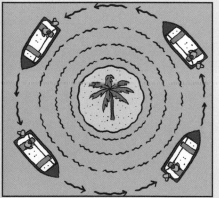

4. But the man hadn't reckoned for the currents. He soon discovers that, although the boat is always pointed on a course that should move it away from the island, it is actually circling it. This is because with each tiny bit of progress the boat makes in a forward direction, the unseen currents pull it in sideways toward the island.

5. The man tries to restart the engine but discovers there is no more fuel. Instead he decides to slow the boat down by blowing into the air in the direction in which the boat is moving.

6. This slows down the boat and, as its forward velocity decreases, the currents spiral it in so that the boat eventually lands on the island. (In a similar way, the Space Shuttle brings itself back to Earth by firing its rockets in a direction opposite to that of its forward motion in orbit).

Space Probes

With the exception of the manned missions to the Moon during the late 1960s and early 1970s, most exploration of the Solar System over the past 50 years has been through the use of space probes—unmanned spacecraft carrying cameras and measuring instruments.

So far, probes have visited all the planets (but not the dwarf planet Pluto, although there is currently a probe on its way), some planetary moons, and several asteroids and comets. Four have left the Solar System altogether. In order to get space probes to distant planets as quickly and economically as possible, frequent usage is made of maneuvers called gravity-assist flybys.

Cassini's Space Mission

The Cassini probe took nearly seven years to reach its target, Saturn, but to get there that fast, it had to be sent on a path that speeded it up by means of gravity assist flybys past Venus (twice), Earth, and Jupiter.

1. Launch from Earth on October 15, 1997. Cassini's trajectory initially takes it inside Earth's orbit, heading for the orbit of Venus.

2. First Venus gravity-assist flyby on April 26, 1998 increases the probe's velocity relative to the Sun.

3. Second Venus gravity-assist flyby on June 24, 1999 further increases its velocity.

4. Earth gravity-assist flyby on August 18, 1999 further increases the probe's velocity relative to the Sun and sets probe on course to Jupiter.

Some Other Notable Unmanned Space Probes

Mission	Launch Date	Destination
Luna 1	Jan. 2, 1959	The Moon
Mariner 9	May 30, 1971	Mars
Voyager 2	Aug. 20, 1977	Jupiter, Saturn, Uranus, Neptune
Giotto	July 2, 1985	Halley's comet
Mars Exploration Rover	June 10 and July 7, 2003	Mars

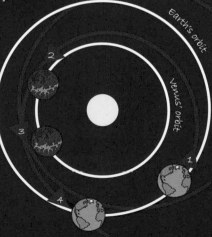

6. Cassini goes into orbit around Saturn on July 1, 2004. It has been orbiting the giant planet ever since. Shortly after arrival, it sent a mini-probe, called Huygens, to Saturn's largest moon, Titan. Huygens landed on Titan's surface and sent back several photographs.

Jupiter's orbit

Saturn's orbit

6

Time Taken	Objective and Results
3½ days to reach the Moon	The first successful space probe, launched by the Soviet Union. It made measurements of the solar wind, flew past the Moon (discovering it has no magnetic field), and went into orbit around the Sun.
5½ months to reach Mars, one year orbiting and photographing surface	The first probe to orbit another planet, it sent back thousands of images of the Martian surface, confirming the presence of huge volcanoes, craters, and canyons.
Nearly 2 years to reach Jupiter, then 2 years to Saturn, 4½ years to Uranus, 3½ years to Neptune	Sent back highly detailed photographs of the four gas giant planets, their moons, and planetary rings. It is now heading out of the Solar System into deep space.
8 months to rendezvous with Halley's comet	Took the first close-up pictures of the nucleus of a comet and successfully made measurements of the cloud of dust and gas surrounding the nucleus.
6 months to reach Mars, over 5 years exploring the surface	Two rover vehicles, called Spirit and Opportunity, have explored different areas of Mars' surface, analyzing its geology and past presence of water.

5. Jupiter gravity-assist flyby December 30, 2000 further increases the velocity and sets probe on course to Saturn. About 26,000 images of Jupiter were taken.

5

FACT: The combined Cassini-Huygens probe weighed as much as a large four-door automobile.

- The theory dates from 1929 following observations by the American astronomer Edwin Hubble.

- It is a key feature of the Big Bang model of the Universe.

- Hubble discovered that distant galaxies are receding from Earth at a rate proportional to their distance.

- The expansion of the Universe is uncontested but there are different theories to explain why it's happening.

The Expansion of the Universe

To understand the expansion of the Universe, you need to grasp the idea of "red shift." Different types of stars—red giants, white dwarfs, and so on—produce mixtures of light and other electromagnetic radiation of different wavelengths. These form a kind of "fingerprint" for that type of star, and can be readily seen by astronomers. Similarly, galaxies also have their own radiation fingerprints. If a particular star or galaxy is moving away from an observer, its radiation fingerprint is slightly different from that of a stationary object of the same type: in fact, all the light and other waves coming from it appear slightly "stretched" because they have lengthened wavelengths. This is called a "red shift," because red light occupies the long-wavelength end of the light spectrum. The greater the object's velocity away from the observer, the greater its red shift.

Relationship to Doppler Effect

Think about a fire truck or ambulance speeding past you on a street. The "Doppler effect" makes the siren pitch get lower as the vehicle moves away from you. And it's the same story with the light waves coming from these stars.

Distant Galaxies Are Getting More Distant

By the early 1920s, it was already known that many stars in our own galaxy, as well as various mysterious, fuzzy-looking objects in the sky, have red shifts—they are moving away from Earth. At this time, telescopes became powerful enough to identify individual stars within the fuzzy-looking objects; it also became possible to estimate accurately the distances to these objects. As a result, American astronomer Edwin Hubble established in 1925 that the fuzzy objects were actually separate galaxies far outside and beyond our own galaxy. Over a number of years, Hubble compared his measurements of the distances from Earth to these remote galaxies with their previously calculated red shifts. The result (now known as Hubble's Law) was quite startling: the farther away the galaxy, the bigger its red shift, and thus the faster it was receding. This applied to scores of galaxies looking in all directions into space. Astronomers quickly realized that if all the galaxies are receding from Earth at a rate proportional to their distance, they must also all be moving away from each other. So in other words, the Universe itself must be expanding.

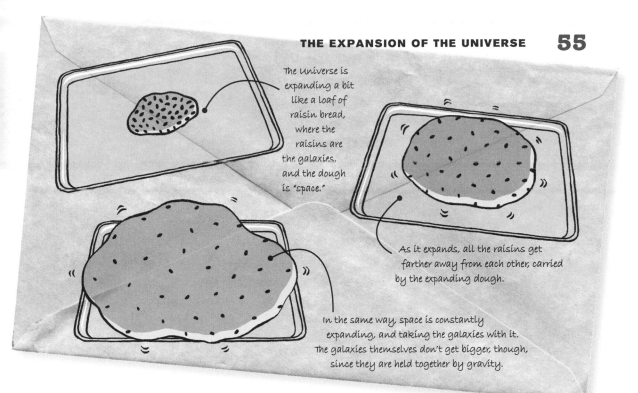

The Universe is expanding a bit like a loaf of raisin bread, where the raisins are the galaxies, and the dough is "space."

As it expands, all the raisins get farther away from each other, carried by the expanding dough.

In the same way, space is constantly expanding, and taking the galaxies with it. The galaxies themselves don't get bigger, though, since they are held together by gravity.

Cue the "Big Bang" Theory

If all the galaxies in the Universe are rushing away from each other, it follows that in the past they must have all been much closer—in a considerably denser (and hotter) Universe than exists today. Thus, Hubble's discovery provided one of the starting points for the development of the "Big Bang" model of the Universe. Since the 1930s, astronomers have continuously been making finer and finer estimates of the rate of expansion (called the Hubble constant), because this can be used to help calculate how long ago the Big Bang occurred.

The expansion only applies over extremely large distance scales. Within localized areas such as individual galaxies, gravity dominates and holds matter together. The effects of gravity would be expected to slow the expansion by attracting galaxies back towards each other. But recent data indicates expansion is actually speeding up. To account for this, scientists are talking about an "anti-gravity" force that helps keep the expansion going. They call this force "dark energy."

The Space Inside Space

The hardest part, conceptually, about the "raisin bread" idea (above) is this. When the loaf expands, it expands into the space we call the oven, in the room we call the kitchen. But when space expands, that's it. There is no pre-existing place into which space is moving. Space is all there is, and everything inside it is getting farther apart.

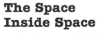

RED SHIFT: Light pattern that indicates that an object is moving away from the observer.

HUBBLE'S LAW: The farther away the galaxy, the bigger its red shift, and so the faster it's receding.

jargon buster

• As yet, no credible evidence for extraterrestrial life has been widely accepted by the mainstream scientific community.

• One objective of a mission to Jupiter's moon Europa, scheduled for 2020, is to look for life in an ocean that may exist below the Moon's surface.

• Four NASA space probes that have left the Solar System carry messages to extraterrestrials in the form of engraved diagrams and audio recordings.

Extraterrestrial Life

The Drake Equation

The Drake Equation was developed by the American astronomer Frank Drake in 1961 as a way of estimating how much intelligent life exists in our own galaxy, based on seven factors. Multiplying these together gives the overall estimate. Some of the factors can only be guessed at, so a range of results is possible. Here's an example based on the author's estimates:

Rate of suitable star formation per year in our galaxy—20

x

Fraction of suitable stars that form planets—0.8

x

Average number of habitable planets per planetary system—0.1

x

Fraction of habitable planets in which life evolves—0.9

x

Probability that life, where it evolves, eventually develops into an intelligent civilization —0.8

x

Probability that an intelligent civilization develops interstellar communication—0.9

x

Average lifespan of these civilizations (years)—5,000

=

Estimated number of civilizations with which we should be able to communicate—5,184

So far, the only life that scientists know about is that found on Earth. However, partly because the Universe is so large, and partly because life as we know it seems capable of thriving in such a wide range of conditions, many are convinced that it must also exist elsewhere. In our galaxy alone, there are hundreds of billions of stars, quite a high proportion of which have planets. Even if only a tiny fraction of the planets have conditions that are favorable to life, that still means the number of potentially habitable planets is large. After that, the question of whether or not extraterrestrial life exists depends largely on how easy or difficult it is for life to begin—whether it can develop only as the product of a highly improbable series of events, or whether it is something that is almost inevitable given a few favorable starting conditions (see pages 68–69).

Would we recognize it if we found it?

Ideas about what life actually is are heavily influenced by the characteristics of life on Earth. However, it is uncertain whether certain characteristics common to all life on Earth, such as a chemistry based on the element carbon, must always be a feature of extraterrestrial life.

Efforts to find extraterrestrial life within the Solar System include sending probes to feasible locations and examining images of planets and their moons for signs of life, so far without success. Outside the Solar System, the main focus is on looking for signals that may have been sent by extraterrestrials. A search has also begun for Earthlike planets orbiting nearby stars. Finally, for over 30 years we have been broadcasting our presence by sending signals toward some nearby stars similar to our Sun. But even if there is life on planets orbiting these stars capable of understanding our messages, we cannot expect a reply from them for several centuries yet.

If and when humans first encounter extraterrestrial life, unfortunately it is far from certain that anyone will recognize or even notice the aliens…

4

Origins Stuff

CHAPTER CONTENTS

This chapter tells the story, according to mainstream science, of how our planet and its inhabitants came into existence. Today, the vast majority of physicists believe that both time and energy (which is the source of matter) originated about 13.7 billion years ago in an event called the Big Bang. The evidence that this happened is now overwhelming, and its details have been worked out down to miniscule time intervals.

After the Big Bang

Starting a relatively short time after the Big Bang, and continuing for billions of years afterward, the atoms that make up our world were forged from simpler components. Gradually, the matter made from these atoms condensed to form galaxies. In a part of our own galaxy, the Solar System formed from what was initially a huge cloud of gas and dust. Within a part of the Solar System, a collision between two young planets led to the formation of the Earth and Moon, and within 600 million years, the first primitive life forms were starting to leave traces of their presence in Earth's oceans. Subsequently, over Earth's long history, the process of evolution led to the appearance of vast numbers of new species, and the extinction of many—as evidenced by the chronicle of life on our planet known as the fossil record.

The Source of Energy

Where did the energy to "fuel" the Big Bang come from? Some physicists think it might have come from absolutely nothing. They argue that when the Universe originated, a vast amount of gravitational potential energy—energy possessed by matter as a result of its position relative to other matter—came into existence. Curiously, this type of energy has a minus value, which means that starting from nothing you can simultaneously get a colossal amount of (positive) mass-energy that turns into matter, and an equally colossal amount of (negative) gravitational potential energy that permeates this matter. Physicists call the process, by which paired positive and negative energy can appear out of nothing, a "quantum fluctuation." Small quantum fluctuations occur all the time in the subatomic world, but whether one big enough to create a Universe like ours is possible remains, at present, an unresolved question.

- At the point when it started inflating rapidly, the temperature of the Universe was about 1,000 trillion trillion°F.

- By the time protons and neutrons had started to form—after about a millionth of a second—it had cooled to about 10 trillion°F.

- Fundamental forces such as gravity and electromagnetism may have been unified as a single force for a short time after the Big Bang.

- It is thought that many particles unknown to today's scientists may have existed for short periods after the Big Bang.

The Big Bang

For much of its very early existence, the Universe consisted of a seething mass of particles and their antiparticles that were forming out of energy and then coming back together again to revert back to energy (a process termed *annihilation*). As it expanded, the Universe cooled. Eventually, temperatures dropped to a point where particles and antiparticles could no longer form out of energy. Most of the existing particles and antiparticles soon annihilated, but for unknown reasons a small excess of particles over antiparticles had built up, and so a residue of free particles remained, including quarks and electrons. The free quark particles combined to form protons and neutrons.

The Universe is now a "soup" of particles in balance with energy (photons). Particles and antiparticles form for short periods from this energy and then annihilate back to energy.

During the first tiny fraction of a second following the Big Bang, the Universe is thought to have undergone an incredibly rapid expansion—by a factor of at least 10^{26} (a 10 followed by 26 zeros). This "inflation" helps explain why the visible Universe is rather uniform in properties such as its density and temperature. Without such rapid inflation, it would have ended up much "lumpier." This phase of the formation of Universe can be likened to the rapid inflation of an airbag in a car—from something small and wrinkled into a much larger, smoother object.

THE FIRST THREE MINUTES OF EXISTENCE:

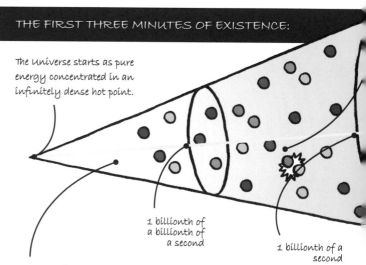

The Universe starts as pure energy concentrated in an infinitely dense hot point.

1 billionth of a billionth of a second

1 billionth of a second

A rapid inflation occurs, with the Universe expanding in a billionth of a billionth of a billionth of a second from a negligible size to about the diameter of a basketball.

Evidence for the Big Bang

Most scientists support the Big Bang theory, convinced by several strands of evidence:

• The fact that the Universe is expanding in all directions implies that in the past it must have been much denser and more compact (see page 54).

• In the 1960s, faint radiation (the cosmic microwave background radiation) was detected that appears to permeate all of space. This radiation can only have come from a uniformly hot, dense object that once filled the whole Universe.

• Mathematical models for how the Universe would have evolved from an extremely hot, dense point show an abundance of elements that exactly match what actually exists.

ANTIPARTICLE: A subatomic particle that has the same mass and average lifespan as the particle to which it corresponds but which has an opposite electric charge. The antiparticle of a quark is an antiquark, and the antiparticle of an electron is a positron (or antielectron).

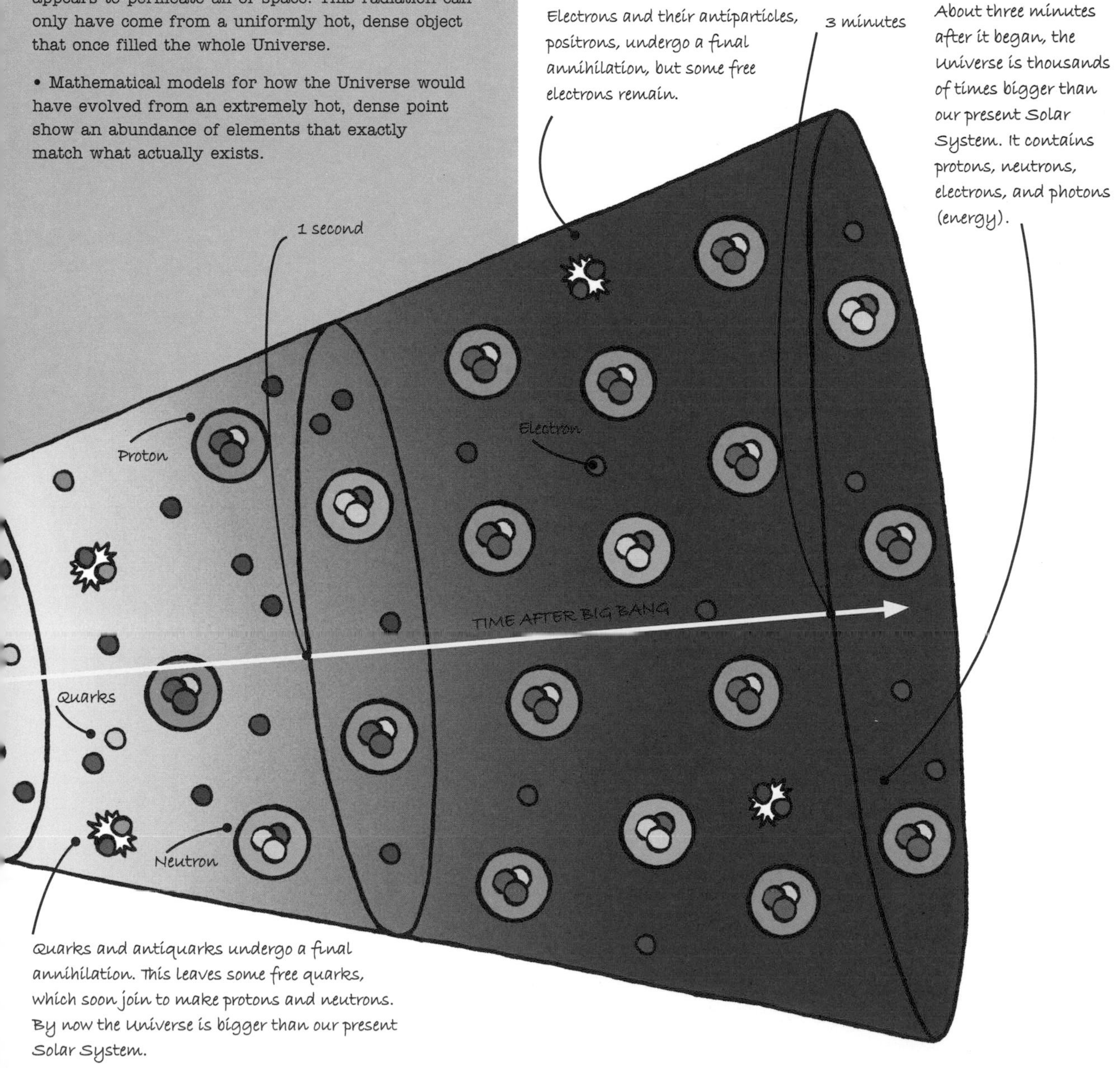

Jargon buster

Electrons and their antiparticles, positrons, undergo a final annihilation, but some free electrons remain.

3 minutes

About three minutes after it began, the Universe is thousands of times bigger than our present Solar System. It contains protons, neutrons, electrons, and photons (energy).

1 second

Proton

Electron

Quarks

TIME AFTER BIG BANG

Neutron

Quarks and antiquarks undergo a final annihilation. This leaves some free quarks, which soon join to make protons and neutrons. By now the Universe is bigger than our present Solar System.

Origins of Atoms and Elements

The first atoms, of hydrogen and helium, originated from events that happened minutes after the Big Bang. Later, the lives and deaths of stars over billions of years enriched the Universe with new chemical elements.

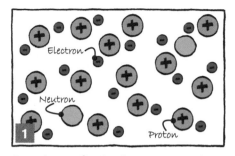

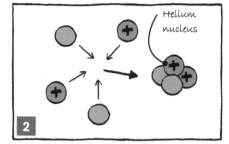

Three minutes after the Big Bang... the Universe contained many free protons and electrons, together with some neutrons, with about seven protons for each neutron.

Over the next 17 minutes... the temperature dropped sufficiently for nearly all the neutrons to combine in pairs with protons, creating the nuclei of helium atoms.

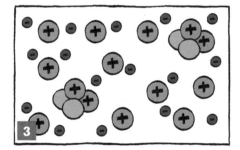

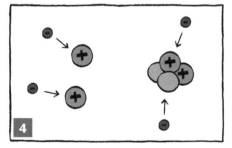

This still left vast numbers of free protons, which outnumbered the helium nuclei 12 to one.

Some 380,000 years later, as the Universe cooled further... the free protons and helium nuclei began to capture electrons to form the first atoms.

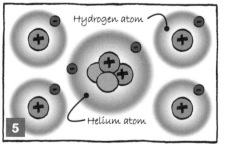

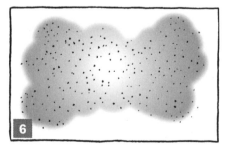

With their captured electrons, the free protons had become hydrogen atoms, while the helium nuclei had become helium atoms.

For a long time afterward (called the "Dark Ages"), the ordinary matter of the Universe was simply a vast expanding cloud of gas, composed almost entirely of hydrogen and helium.

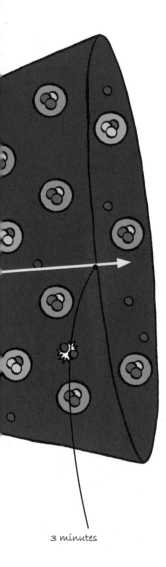

3 minutes

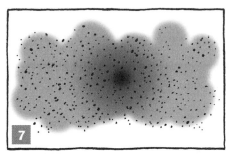

7

After a few hundred million years... gravity created high-density regions in this gas cloud that began to form intricate structures: The first galaxies.

8

Gas clumps within the galaxies began to contract to form stars. Within the largest stars, the atomic nuclei of new elements, such as oxygen and iron, were formed.

Jargon buster

HELIUM: A chemical element that normally consists of a low-density, unreactive gas consisting simply of single helium atoms. It is the second most common element in the visible Universe (after hydrogen) because so many helium nuclei were made in the first 20 minutes after the Big Bang.

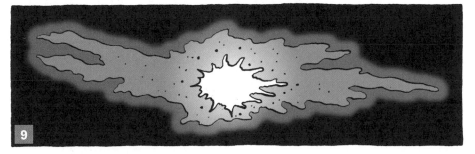

9

The largest of these primordial stars exploded at the end of their short lives in supernovas (see pages 46-47), creating yet more chemical elements, such as gold and uranium.

10

The supernova explosions, and the quieter deaths and disintegration of smaller stars, dispersed the atoms of the new chemical elements into the vast clouds of hydrogen and helium that still make up most of the ordinary matter in galaxies. Initial chemical reactions between atoms in these clouds eventually led to fine particles of dust forming. In this way, the scene was set for the next stage in the process—the formation of stars like our own Sun, with encircling planetary systems.

- The Solar System formed between about 4.7 and 4.54 billion years ago from a gigantic cloud of gas and dust.

- Our Sun ignited as a star when the temperature at the center of this contracting cloud reached 22 million °F.

- The same type of process that formed the Sun and planets has now been observed occurring elsewhere in our galaxy.

- It has been suggested, though not proven, that a shock wave from a supernova may have triggered the process that led to the formation of the Solar System.

Origin of the Solar System

The first galaxies are believed to have started to form around 500 million years after the Big Bang. Over the subsequent several billion years, each galaxy was gradually enriched with new chemical elements formed in stars and supernovae. Within our own galaxy, the Milky Way, about 4.7 billion years ago a huge cloud of material, perhaps as much as a light-year across, began to condense into what was to become the Solar System. This cloud, called the solar nebula, consisted mainly of hydrogen and helium gas, but it also held some dust containing metallic elements and particles of ice made of substances such as water and methane. The basic process by which this cloud turned into the Sun and planets is described below.

Milky Way Nebulae

To astronomers, a nebula is any cloud of dust and gas occupying an otherwise empty region of the Milky Way. Some large nebulae contain regions where new stars are being born. Others are the remnants of supernovas or other star deaths (see pages 46–47).

How the Sun and Planets Formed

The most widely accepted theory for how the Solar System formed, called the solar nebula hypothesis, is shown here. It provides a plausible explanation for many of the basic facts about the Solar System, for example, why the orbits of the planets around the Sun all lie in the same plane, and why they all orbit in the same direction. The whole process, from the collapse of the solar nebula to the establishment of the planets in orbit around the Sun, is

1. A huge, slowly spinning cloud of cold gas, dust, and ice particles, several times the mass of the Sun today, began to collapse (condense) under the influence of gravity.

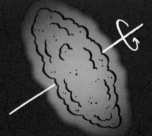

2. Just as an ice skater spins faster as she pulls in her arms, the disk rotated faster as it shrank. It also started to flatten out. A denser area began to form at its center, and this area gradually warmed up.

The Origins of Orbiting Bodies

Four main types of objects orbit the Sun—the inner rocky planets, the outer gas giant planets, small rocky bodies called asteroids, and in the outer parts of the Solar System, small objects made of ice and rock, including comets. The solar nebular hypothesis (below) explains how each of these may have formed from different parts of the protoplanetary disk:

• Close to the protosun, temperatures were so high that only rocky particles and metals could remain in solid form. Other materials were vaporized, and most of the gas in this region was eventually dissipated by the Sun's radiation. In due course the rocky and metallic particles gave rise to the inner planets, including the Earth.

• In the cooler, outer regions of the disk, the solid particles contained various types of ice as well as rock. Most of the planetesimals that formed here eventually came together to form four large objects made of ice and rock. These bodies, which were to become the gas giant planets, were massive enough to attract vast amounts of various gases into the thick atmospheres that formed around them.

• The remaining planetesimals in the outer parts of the disk gave rise to comets and other small bodies made of ice and rock.

• A ring of planetesimals in the middle part of the disk never aggregated to form a planet, possibly because of the disruptive gravitational influence of Jupiter. Instead, this ring became the asteroid belt.

The Age of the Solar System

The Solar System is thought to have finished forming between about 4.57 and 4.54 billion years ago. How do scientists know this? The main evidence comes from meteorites—rocks that have reached Earth from outer space. Some meteorites, called chondrites, are generally recognized to be leftover debris from the formation of the Solar System. Many of them have been carefully dated, and they are all about 4.56 billion years old. This strongly suggests that the final stage of Solar System formation occurred around this time and that it happened relatively quickly. If this were not the case, there would be considerably more variation in the age of the chondrites.

PLANETESIMAL: Any of innumerable small objects made of rock or rock and ice that orbited within the spinning disk of material from which the Solar System was formed.

METEORITE: A stony or metallic object that has fallen to Earth's surface from space. Some meteorites are thought to be planetesimals that never joined up with others to form planets. Others are fragments of larger objects that have broken up, or are chunks of material from Mars or the Moon, ejected as a result of past impacts.

Jargon buster

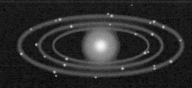

3. The cloud eventually flattened into a pancake shape, with a central bulge surrounded by a thin disk (the protoplanetary disk). Within the disk, grains of dust and ice began to collide and stick together. Over tens of millions of years, these particles grew to form larger objects, made of rock and/or ice, known as planetesimals.

4. Once the planetesimals had grown to a few miles in diameter, their gravity began to sweep up more and more dust and ice, in an escalating process. Meanwhile, the central bulge had condensed further into an increasingly dense hot object known as the protosun.

5. Some 4.57 billion years ago, the protosun became the Sun, a star. It began to radiate light and heat. By this time, most of the planetesimals had coalesced to form large bodies called protoplanets. About 4.54 billion years ago, these underwent a series of violent mergers to form the inner planets and the cores of the outer planets.

bite size facts

• Soon after the collision that is thought to have formed them both, the Earth and the Moon were rotating much faster than they are today. An Earth day may have lasted as little as 5 hours.

• There are no surface rocks left on Earth from its original formation, since all of its original surface has melted again at least once.

• The oldest Moon rocks are older than the oldest Earth rocks. This is because parts of the Moon's surface appear to have solidified soon after its formation and have never melted again.

Origin of Earth and the Moon

Our planet formed from the gradual accumulation and condensation of particles of dust, and later of larger rocky and metallic particles, in a disk of material rotating around what was to become the Sun (see pages 64–65). By the end of this process, about 4.54 billion years ago, a fiery-hot ball of liquid rock and metal, the proto-Earth, came into existence. Most planetary scientists now believe that about 10 to 50 million years after its original formation, another young planet, which has been named Theia, crashed into the proto-Earth. The debris spewed out from the collision later coalesced to form the Moon.

Where Did Theia Come From?

According to one theory, Theia formed at the same time as the proto-Earth and in the same orbit around the Sun, but at a considerable distance from the proto-Earth. For unknown reasons, at some point Theia became destabilized in this orbit leading to its collision with the proto-Earth.

How the Proto-Earth Formed

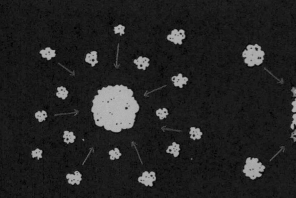

1. Within the disk of gas and dust that surrounded the protosun in what was to become the Solar System, dust grains accreted (gathered together) by collisions and gravitational attraction to form ever larger particles (planetesimals).

2. The larger planetesimals grew fastest, and once these had reached a few miles in diameter, their greater gravitational attraction caused collisions to occur with increasing violence and frequency.

3. The collisions caused a huge amount of heat energy to be released, and by the time the last of the collisions had occurred—between the now large objects called protoplanets—the proto-Earth was a fiery globe of molten rock.

Birth of the Moon

Many different theories have been put forward for how the Moon formed, but only the one outlined here is widely accepted today. One of the main arguments supporting this theory is that it accounts for why the Moon has only a small iron core, whereas Earth itself has a large iron core. According to this theory, the Moon formed from the outer layers of Theia and the proto-Earth, while most of the iron cores of the two protoplanets merged to create what today is Earth's core.

1. Between 10 and 50 million years after its own formation, the proto-Earth was hit by another young planet about the size of Mars—Theia.

2. A huge amount of material from the impacting bodies' outer, lighter layers was spewed out. Meanwhile, their heavier, iron-rich cores merged.

3. The ejected material formed a ring of partially molten rocky debris that encircled Earth. Particles within this ring accreted to form larger particles, which began to sweep up the rest.

4. Possibly within as little as a few months, all the material in orbit had come together to form the Moon. At this point both the Earth and Moon were molten. The outer crust of the Moon and later that of Earth subsequently solidified, and the two bodies have been cooling ever since.

Consequences of the Impact

The cataclysmic impact between Theia and the proto-Earth, together with its result, the Moon, had some profound effects on Earth and its subsequent history. For a start, the collision is suspected to have knocked the proto-Earth out of position, producing the tilt in its spin axis, which is the cause of seasons. However, the Moon's presence has also helped stabilize this tilt, which is likely to have smoothed out climatic variations over the whole course of Earth's history.

Other important effects are likely to have been exerted through forces produced by the Moon's gravity at Earth's surface. When it first formed, the Moon was closer to Earth than it is today, and these forces were more pronounced. Their strength may have triggered the development of Earth's system of tectonic plates (see pages 84–85). They also produced tides in Earth's oceans (see pages 90–91). By subjecting organisms that lived in tidal zones to daily (or twice-daily) fluctuations in their living conditions, the Moon may have influenced the pace at which life on Earth evolved.

ACCRETION: The joining together of small objects into a larger object by gravitational attraction and low-speed collisions, causing the objects to stick to each other.

Jargon buster

• The oldest traces of life on Earth, detectable in some ancient rocks and rock formations, date from around 3.5 billion years ago.

• The most widely accepted theory is that life developed in Earth's oceans from chemical building blocks formed in its early atmosphere.

• Between about 4.5 and 3.5 billion years ago, Earth was probably hit by many comets. It is possible that these brought the building blocks of life to Earth from space.

Origin of Life on Earth

Over the past 100 years or so, many different explanations have been put forward for how life on Earth might have arisen from non-living matter. One characteristic of all known life is that it utilizes certain organic (carbon-containing) substances with large and complex molecules, such as proteins and nucleic acids. Wherever life originated, the basic chemical building blocks for these substances, or simpler forms of them, must have come first. The best-accepted theory is that these chemical building blocks formed through reactions between a few simple gases present in Earth's early atmosphere. These substances then washed down into the oceans, where they combined to form more complex substances.

1 Primitive Gases

Earth's early atmosphere is likely to have contained a number of simple gases, such as hydrogen, methane, ammonia, and water vapor, although no free oxygen. Discharges of lightning in the atmosphere may have supplied the energy for molecules of these gases to combine into slightly more complex substances, such as amino acids.

2 Chemical Soup

Once amino acids (and possibly various other carbon-containing substances) had formed in the atmosphere, they are thought to have washed down into Earth's oceans, where they combined in a "chemical soup" to form the more complex molecules characteristic of living things.

NUCLEIC ACID: A complex substance found in living things that forms the basis of genes, the coded instructions for how living organisms work. DNA is one type of nucleic acid.

3 Rise of the Replicators

For life to have originated from these complex molecules, some type of self-replicating molecule must have developed, as a result of billions of random interactions between non-replicating molecules. This molecule would have been the earliest ancestor of DNA, of which genes—the coded instructions by which living organisms today operate (see pages 146–147)—are composed.

4 Early-Life Chemistry

Another precondition for life is that it must have developed a means of extracting energy from its environment. Some biologists think that life is most likely to have first developed on the floor of the oceans, near hydrothermal vents or "black smokers" that poured vast amounts of heat energy into the oceans, as well as energy-containing chemicals such as methane. A later development—and an extremely important step for the further evolution of life—was the appearance of organisms that can extract energy from sunlight through the process called photosynthesis.

Electrode

3. To simulate lightning, a high-voltage electrical power source was attached across the electrodes, repeatedly subjecting the gas mixture to sparking.

Cooling apparatus containing flowing water

The Miller-Urey Experiment

This experiment, first performed in the 1950s, attempted to simulate conditions in Earth's early atmosphere. It was successful in showing that these conditions could have resulted in the formation of amino acids, the building blocks for protein molecules, which are utilized by all known life forms.

2. A mixture of the simple gases likely to have been present in Earth's early atmosphere was introduced into the apparatus.

Might Life Have Come from Space?

Some scientists think that primitive life, or at least complex biological molecules, originally formed in space or on a nearby planet, such as Mars, and was brought to Earth on a comet or meteorite. Some support for this idea has come from studies using a technique called spectroscopy (identifying chemical compounds remotely by analyzing the light they emit). This has shown that various complex organic (carbon-containing) molecules, the building blocks for life, exist in space, and has also suggested that the surfaces of comets may contain a mixture of such substances. However, so far, nothing identifiable as life has been detected beyond Earth. If and when it is, the discovery would still not answer the question of how life first originated.

1. Some water was heated up in a large flask

4. Analysis of the condensate showed that it contained various amino acids.

- If Earth's whole history was compressed into one year, then each day would represent 12.5 million years.

- According to this calendar, animals and plants didn't evolve on Earth until the late fall.

- Fish first started crawling onto land, evolving into amphibious creatures with legs, in early December.

- Primates—the animal group that today includes lemurs, monkeys, humans, and other apes—only appeared in the last week of December.

Earth's History

In the 4.54 billion years since it was first formed, our planet has undergone some incredible upheavals, ranging from long periods when it was heavily bombarded by asteroids and comets to interludes when it was completely covered in ice. For about three-quarters of this time, life has been present on Earth. The various geological processes and events that have occurred have been closely intertwined with the evolution of life. For instance, some types of living things only became possible when changes in the chemistry of the atmosphere and oceans allowed them to thrive and evolve. In turn, the lives and deaths of organisms have changed Earth's chemistry—for example, by oxygenating the atmosphere or by leaving thick deposits of calcium carbonate on the sea floor. Catastrophic geological events such as asteroid impacts and major volcanic eruptions have had a significant effect on the evolution of life, notably by causing mass extinctions.

If the whole of Earth's history is represented on a one-year calendar (see right), then each month represents about 380 million years. For some of the months, particularly December, a huge amount is known, all gleaned from rocks and fossils. Other months (such as April and May) are "dark ages," not necessarily because nothing happened during these times, but because most of the evidence of what did happen has been destroyed.

STROMATOLITE: A layered clump of rock, made from the mineral secretions of microscopic living organisms. Some stromatolites are modern; others (fossils) were made billions of years ago.

	January	February	March
1	1	1	Formation of proto-Earth
2	2	2	2
3	3	3	3
4	4	Formation of Earth–Moon system following collision between proto-Earth and another young planet	4
5	5		5
6	6		6
7	7	7	7
8	Oldest-surviving mineral deposits form	8	Oldest-surviving rock forms
9	9	9	9
10	10	10	10
11	11	11	11
12	12	12	12
13	13	13	13
14	14	14	First simple life-forms evolve in oceans
15	15	15	
16	16	16	
17	17	17	17
18	18	18	18
19	19	19	19
20	First ocean forms	20	20
21	21	21	21
22	22	22	22
23	23	23	23
24	24	24	24
25	25	25	25
26	26	26	26
27	27	27	27
28	28	28	First stromatolites—rocky masses made by microscopic organisms
29	29	29	
30	30	30	
31	31		31

The History of the Earth on a 12-Month Calendar

April	May	June	July	August	September	October	November	December
1	1	1	1	1	1	1	1	1
2	2	2	2	2	2	2	2	2
3	3	3	3	3	3	3	3	3
4	4	4	4	4	4	4	4	4
5	5	5	5	5	5	5	5	5
6	6	6	6	6	6	6	6	6
7	7	7	7	7	7	7	7	7
8	8	8	8	8	8	8	8	8
9	9	9	9	9	9	9	9	9
10	10	10	10	10	10	10	10	10
11	11	11	11	11	11	11	11	11
12	12	12	12	12	12	12	12	12
13	13	13	13	13	13	13	13	13
14	14	14	14	14	14	14	14	14
15	15	15	15	15	15	15	15	15
16	16	16	16	16	16	16	16	16
17	17	17	17	17	17	17	17	17
18	18	18	18	18	18	18	18	18
19	19	19	19	19	19	19	19	19
20	20	20	20	20	20	20	20	20
21	21	21	21	21	21	21	21	21
22	22	22	22	22	22	22	22	22
23	23	23	23	23	23	23	23	23
24	24	24	24	24	24	24	24	24
25	25	25	25	25	25	25	25	25
26	26	26	26	26	26	26	26	26
27	27	27	27	27	27	27	27	27
28	28	28	28	28	28	28	28	28
29	29	29	29	29	29	29	29	29
30	30	30	30	30	30	30	30	30
	31		31		31			31

Short-term rise in oxygen levels caused by first photosynthetic organisms

Microscopic organisms that left the first microfossils living in an area of what is now South Africa

Asteroid hits what is now South Africa, leaving a 170-mile-wide impact crater

Dinosaurs are the dominant life-form on Earth

First jawed fish appear in the oceans

The Himalayan mountain range begins to form

Mass extinction, possibly caused by massive volcanic eruptions. Large numbers of species are wiped out.

First mammals evolve

First animals with hard body parts appear in the oceans

Large meteorite hits Earth, probably contributing to the extinction of the dinosaurs

Modern humans evolve

First tree-like plants evolve on the land

The Last 15 Minutes

If all of Earth's history is represented on a 12-month calendar, then the last 15 minutes on this calendar (from 11:45 P.M. to 12 midnight on December 31) represents the last 130,000 years. On this basis, each minute represents 8,630 years and each second represents about 144 years. A few "days" earlier in the history of Earth, the dinosaurs had become extinct, although they left a group of descendants that we know today as birds. And just a few minutes earlier, modern humans had evolved, almost certainly somewhere in Africa. Some of the most notable events during this last 15 minutes have been an ice age extending throughout most of the period, a big drop and later a rise in sea level, the spread of modern humans across the world, and the extinction of many large mammals, quite possibly as a result of human activity.

Two Minutes and Counting...

11:58 20,000–14,000 years ago: Anatomically modern humans first reach the Americas.

11:58 18,000 years ago: Maximum extent of ice of last ice age. Ice-sheets extend deeply into North America, Europe, and Siberia. South of the ice-sheets, mammoths, giant deer, horses, and wooly rhinos graze.

11:58:30 12,500–12,000 years ago: End of last ice age. Ice-sheets in full retreat; many continental shelf areas such as the North Sea are flooded as the ice melts and sea levels rise.

11:58:40 12,000 years ago: First settled agriculture in Turkey, Middle East, and Mesopotamia.

11:58:40–11:59:25 12,000–5,000 years ago: Many large mammals—for example, wooly mammoths, saber-toothed cats, giant ground sloths, and cave bears—become extinct, possibly as a result of hunting by modern humans.

11:58:45 c. 11,500 years ago: Beginning of Neolithic Age (New Stone Age) in Middle East.

11:59 c. 10,000 years ago: Recorded human history begins.

11:59:10 8,200 years ago: Major Earth cooling lasting for several centuries.

11:59:20 c. 5,500 years ago: First city-states in Mesopotamia.

11:59:30 c. 4,500 years ago: Great Pyramid of Giza built.

11:59:40 c. 3,000 years ago: Beginning of Iron Age in Middle East.

11:59:50–59 400–150 years ago: Little Ice Age, a cooler period in many parts of the world.

A VERY BRIEF HISTORY OF THE LAST 130,000 YEARS

11:52 74,000 YEARS AGO: HUMANS ALMOST WIPED OUT BY VOLCANIC ERUPTION
Massive eruption of Toba supervolcano in Sumatra blasts huge amounts of ash into the atmosphere, causing a big drop in global temperatures for several years.

11:50–11:53 85,000–60,000 YEARS AGO: EARLY MODERN HUMANS MIGRATE FROM AFRICA

11:47 115,000 YEARS AGO: ICE AGE STARTS
The most recent ice age, which turns out to last for more than 100,000 years, begins. Sea levels drop.

11:45–11:47 130,000–114,000 YEARS AGO: FORESTS IN THE ARCTIC
A warm period on Earth is sandwiched between ice ages. Global sea level is 13–20 feet higher than today. Forests reach as far as the Arctic Circle. Anatomically modern humans are present only in Africa.

11:54–11:56 55,000–30,000 YEARS AGO: HUMAN MIGRATION CONTINUES Groups of early-modern humans spread to South Asia, Australia, Europe, and East Asia (in that order); they are hunter-gatherers.

11:57–11:59 24,000 YEARS AGO: EXTINCTION OF NEANDERTHALS (relatives of modern humans).

11:59:59 1883: KRAKATOA ERUPTS Eruption of the Indonesiam volcano causes up to 230,000 human fatalities as a result of ash falls and tsunamis.

11:56 32,000–10,000 YEARS AGO: HUMAN CREATIVITY DEVELOPS Examples of cave art (by early-modern humans) in Europe.

11:59:25 5,000 YEARS AGO: INDIAN OCEAN CATACLYSM A large meteorite hits the Indian Ocean and causes gigantic tsunamis that may be the origin of various flood stories.

- Darwinian evolution is a change in a species, brought about at least partly by a mechanism called natural selection. It can ultimately lead to the development of new species.

- Natural selection is the main mechanism for "adaptive evolution"—the process by which living things can become better adjusted to their environment.

- Other processes involved in evolution include the production of variation in a species, competition for resources between the members of a species, reproduction, and inheritance.

Darwinian Evolution

Our planet supports a fantastic variety of life forms, each of which is astonishingly complex. In the 19th century, the English naturalist Charles Darwin developed and published a theory explaining how all life on Earth developed from common ancestors through a process called evolution by natural selection. In conjunction with a source of variation in a population of organisms—which arises through changes called *mutations* in their genetic make-up—natural selection explains how and why populations of organisms can evolve over time. Alongside circumstances that will sometimes isolate groups of organisms from other populations of the same species, it also explains how new species can arise. Finally, the same theory successfully explains how complex life forms can develop from simpler ones over very long periods of time.

From Variation to Speciation

ISOLATION

Suppose a mouse population lives on a continent that subsequently divides in two, with one part moving north and the other south. The mice are split into two populations, which are now isolated from each other, without the possibility of interbreeding. Suppose, also, that environmental conditions on the two continents start to change quite markedly, in different directions.

SPECIATION

Tens of thousands of generations later, largely as a result of mutations and natural selection, the two mouse populations have evolved differently to suit what are now quite different environmental conditions. The mice on the northern island are larger, have relatively short tails, and brown fur with spots. On the southern island, they have all-white fur, extra-long tails, and orange ears. The rodents from the two islands are unable to breed together—they have diverged sufficiently in their genetic make-up that they are now separate species.

Jargon buster

SPECIES: A group of living organisms that are different from all other groups of organisms and whose members can interbreed, producing fertile offspring.

Natural Selection, Evolution, and Speciation

By following the story of a group of mice in a particular geographical region, we can see how natural selection—the key process in evolution—actually happens. Combined with the geographical isolation of two groups of mice (see panel, left), it can lead to speciation, the appearance of new species.

1. VARIATION

A population of mice come with three different coat colors (in equal proportions) and with two different tail sizes. This variation among the mice is the result of mutations that occurred in the genetic material of their ancestors. Three random mice pairs from this population...

2. COMPETITION

...have a large number of offspring. These young mice will need to avoid being killed by predators if they are to reach adulthood and reproduce. Their ability to survive and reproduce also depends on being able to compete successfully for food and mates.

3. SELECTION

Selection is like a sieve that tends to screen out some characteristics but lets others through. These characteristics can be anything that affects the chances of an organism reaching adulthood and reproducing.

4. SURVIVORS OF SELECTION

Only a certain percentage of the mice make it through the selection sieve to reach adulthood, where they are now ready to reproduce. In this case, predators have found it easier to catch white or long-tailed mice than dark-colored mice or mice with short tails. Dark colors and short tails have been "selected for," while white fur and long tails have been "selected against."

5. INHERITANCE

Through the mechanisms of inheritance, the next generation of mice contains a far higher proportion of dark-furred and short-tailed individuals than the original population. Thus, the mouse population has already undergone a significant change. Over many generations, this process, combined with occasional mutations, can lead to further significant change in a population of living organisms, both at the level of the hereditary material that is passed on from one generation to the next—the genes in that population—and in the outwardly observable properties of the mouse population. This change is what is meant by "evolution."

- Fossils range from microscopic traces of bacterial cells in rocks to gigantic dinosaur bones and petrified tree trunks.

- An estimated 99.9 percent of all species of organisms that have ever lived are now extinct, existing today only as fossils, if at all.

- Hundreds of thousands of extinct species of animals, plants, and other organisms are now known to us by their fossils.

- As well as the preserved bodies or body parts of animals, there are fossils of other remains such as animal footprints and feces.

The Fossil Record

The fossil record is the history of life on Earth as revealed by the remains of plants, animals, and other organisms preserved in rocks. Fossilization can occur in various ways—for example, by mineral replacement of the bones of a dead animal while these are buried deep underground. However, it is a rare process, so probably only a tiny fraction of living organisms that have ever lived on Earth have left fossils.

Fossils and Evolution

Fossils have been key to helping scientists unravel how life on Earth has evolved. In the 19th century it was noticed that different rock layers contain different groups of fossils, most of which are of species that no longer exist. This led to the realization that animals and plants can become extinct. At the same time, it became clear that Earth has been through different ages, represented by different rock layers. The fossils in the rock layers provide evidence for what life forms existed in each age, and also give clues to what living conditions were like on Earth at the time.

Some fossils from different rock layers are clearly of similar types of animals or plants, whether sea urchins, ferns, dinosaurs, or horses. Looking at a particular group, progressive changes can often be seen, going from the fossils in older rock layers to those in more recent layers. Studies of this type continue to be extremely important in mapping the past course of evolution.

Human Fossils and the "Recent African Origin" Hypothesis

Since the 19th century, the fossilized skeletal remains of dozens of hominids (relatives of modern humans) have been unearthed in various parts of the world. All those more than about two million years old come exclusively from Africa. More recent ones, between two million and 30,000 years old, have been found in various parts of Africa, Asia, and Europe. This suggests that a first wave of hominids migrated out of Africa around two million years ago. However, other studies have strongly suggested that modern humans evolved in Africa between about 200,000 and 100,000 years ago, and it wasn't until around 65,000 years ago that the first small groups of them migrated out of Africa. These bands of humans multiplied, and then fairly quickly spread in waves to all other parts of the world, displacing other hominid species, which died out. Strong evidence for this theory comes from analysis and comparison of the DNA (genetic material) of people from many different parts of the world (see pages 150–151).

How a Fossil Forms

A dead animal or plant has to go through a long and complex process before it can reappear at Earth's surface as a fossil. It's only rarely that all the conditions are right for this process to happen.

1. First the organism must die in a place where it will soon be covered in sediment such as sand or silt. Typical environments include the bottom of a lake or mudflats.

2. Some of the dead organism's soft parts may be eaten by surface scavengers. The rest are consumed by small organisms, or slowly decay once the body is buried, leaving behind the hard parts such as teeth and bones.

3. As new layers of sediment settle, the skeleton is buried deep below the surface. Pressure causes particles of sediment to fuse together, forming a sedimentary rock, which encases the skeleton.

4. Groundwater laden with minerals seeps around the bones. Minerals are left in the spaces within the bone and may partially or completely replace the bone itself.

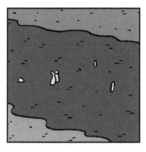

5. For the fossil to be discovered, the rock layer it lies in has to be lifted up by the Earth's movements, and the layers above it eroded away by action of water, wind, or ice.

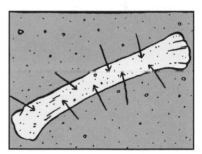

• Commonly used classification schemes divide all life into either five or six kingdoms.

• Kingdoms contain subgroups called phyla, and below phyla there are further subgroupings such as classes, orders, and species.

• If a complete tree of life were to be drawn, it would have tens of millions of endpoints, each representing a species of organism.

• The number of discovered and undiscovered living species of insects is estimated at between 6 and 10 million.

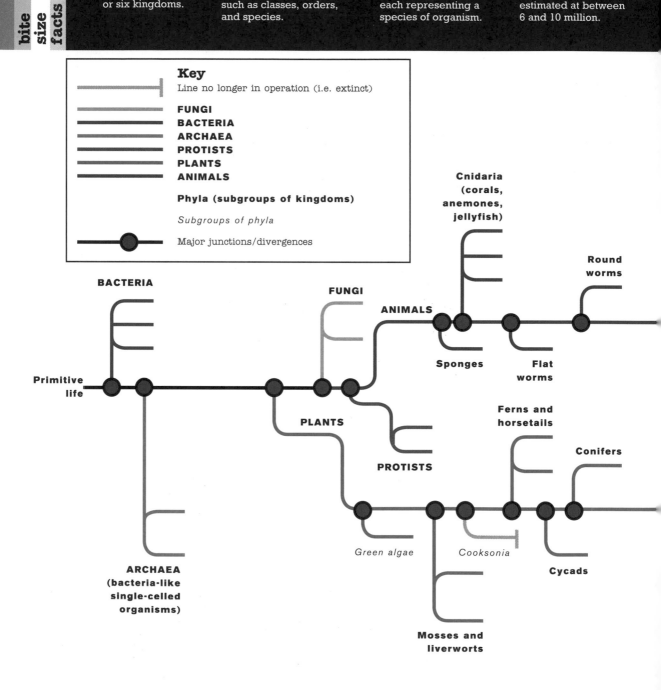

Key

Line no longer in operation (i.e. extinct)

FUNGI
BACTERIA
ARCHAEA
PROTISTS
PLANTS
ANIMALS

Phyla (subgroups of kingdoms)

Subgroups of phyla

Major junctions/divergences

BACTERIA

FUNGI

ANIMALS

Cnidaria (corals, anemones, jellyfish)

Round worms

Sponges

Flat worms

Primitive life

PLANTS

PROTISTS

Ferns and horsetails

Conifers

Green algae

Cooksonia

Cycads

ARCHAEA (bacteria-like single-celled organisms)

Mosses and liverworts

Tree of Life

The study of the structures of living organisms and fossils has allowed scientists to work out in detail how they are related to each other. This knowledge has been gathered over many decades, and new life forms are continually being discovered—both living and extinct. If all the relationships are laid out as a big diagram, the evolution of life on Earth can be seen to have been like the growth of a railway network, with new branch lines splitting off and diversifying by forming sub-branches. Biologists now recognize three main categories or "domains" of life forming this network. The first two of these—bacteria and archaea—consist of relatively simple cells that have no nuclei, called *prokaryotes*. The third domain, which branched off from the others more than two billion years ago, consists of the *eukaryotes*—their cells are more complex and have nuclei. The eukaryotes diversified into such groups (called kingdoms) as plants and animals. These branch into subgroups (or phyla) such as (among animals) the mollusks and arthropods.

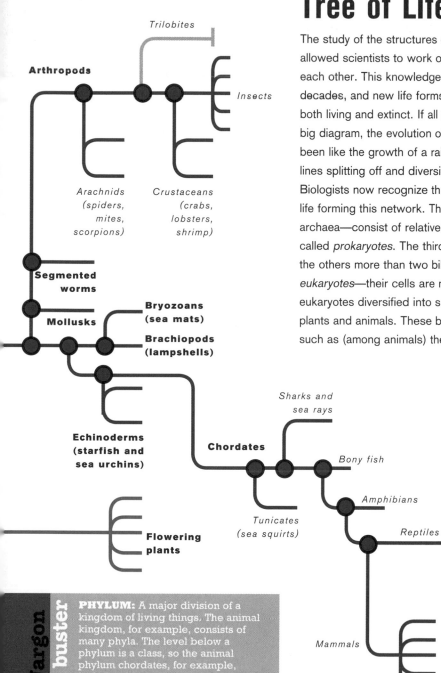

Trilobites

Arthropods

Insects

Arachnids
(spiders,
mites,
scorpions)

Crustaceans
(crabs,
lobsters,
shrimp)

**Segmented
worms**

**Bryozoans
(sea mats)**

Mollusks

**Brachiopods
(lampshells)**

**Echinoderms
(starfish and
sea urchins)**

Chordates

Sharks and
sea rays

Bony fish

Bird-like
dinosaurs

Birds

Amphibians

Other
dinosaurs

Reptiles

**Flowering
plants**

Tunicates
(sea squirts)

Other reptiles

Mammals

Primates (including humans)

Rodents

Cetaceans

Bats

Others

**Jargon
buster**

PHYLUM: A major division of a kingdom of living things. The animal kingdom, for example, consists of many phyla. The level below a phylum is a class, so the animal phylum chordates, for example, includes the classes mammals, amphibians, and reptiles.

5

Earth Stuff

CHAPTER CONTENTS

Our world is a restless planet with a dynamic behavior driven by two main sources of energy—the production of colossal amounts of heat in its interior, and the vast quantities of energy poured out from the Sun. This chapter is an overview of the workings of our planet.

Interior Forces

The energy sources from Earth's interior originate from the decay of radioactive isotopes in rock layers deep within the planet. The released heat causes vast churning movements of these rocks, allowing heat to be carried outward. As a result of this activity, chunks of Earth's outermost shell, known as tectonic plates, are moved slowly around over the surface of the planet. These plate movements not only account for why continents such as Africa, South America, Antarctica, and Australia—once joined in a single "supercontinent"—are now separated by vast oceans, they also help explain a large number of other geological phenomena, such as how mountains form, what causes earthquakes and volcanic activity, and the reason for the extensive systems of ridges and deep trenches that have been discovered on the floors of the world's oceans. Bit by bit, the plate movements also rework and recycle the rocks of Earth's outermost layer, or crust.

The Impact of Solar Energy

The Sun's energy, impinging on Earth's atmosphere and surface waters, produces phenomena such as winds, ocean currents, and water evaporation from the sea surface. Solar energy has a controlling influence on the climate and weather, as well as on Earth's water cycle, which in turn drive processes such as weathering, erosion, and the deposition of sediments that continuously reshape the planet's land surface. The workings of Earth's ocean currents and daily tides are examined in this chapter, as well as such newsworthy phenomena as the existence of a large plastic "garbage patch" in the middle of the Pacific Ocean and the current rate of sea-level rise. Also included are descriptions of the structure of Earth's atmosphere, the causes of good and bad weather (along with explanations of what terms such as warm and cold "fronts" actually mean), and the curious but impressive phenomenon of lightning.

- Earth has three main layers: the crust (on the outside), the core (at the center), and between these two, a thick intermediate layer, the mantle.

- Most of our knowledge about the interior of Earth has come from information revealed by the study of earthquake waves.

- About 99 percent of the rock in Earth's interior is hotter than 2,000°F.

- The processes of radioactive decay generate heat in Earth's interior faster than the current rate of human global energy consumption.

Earth's Structure

When Earth first formed, our planet was a molten (liquid) or semimolten sphere for a few million years. While it was molten, gravity concentrated denser material near the center and less dense materials near the surface. As Earth solidified, it developed an internal structure of three main layers: a thin crust of relatively light rocks; a much deeper layer of denser rocks—the mantle; and at the center, a heavy core, made of iron and nickel.

GEOTHERMAL: Relating to heat generated in Earth's interior by the decay of radioactive elements such as uranium and thorium.

Jargon buster

CORE
This occupies 54 percent of the total diameter, similar to the proportion taken up by an egg yolk. The temperature at its center is around 10,000°F.

MANTLE
This accounts for about 45 percent of the diameter, similar to the proportion of white in an egg. The temperature in its deepest part is around 7,000°F.

CRUST
This is 7–43 miles in thickness—0.2–1.1 percent of Earth's total diameter. The temperature at its boundary with the mantle is between 700 and 1,500°F.

Earth as an Egg
The proportions of crust, mantle, and core that comprise Earth are similar to the proportions of shell, white, and yolk in an egg.

Evidence for Earth's Core

It's easy to discover that an egg has a yolk, but how do scientists known that Earth has an iron core? Most knowledge about Earth's internal structure has come from analysis of earthquake waves. Some of these waves can travel right through the planet and are detectable where they arrive back at the surface. In the 20th century, scientists calculated from the pattern of wave arrivals that there must be a particularly dense region at the center of Earth. The size and density of this core was estimated, and it was soon realized that the core was probably made mostly of iron. This fitted with the density estimate and the presence of iron also explained Earth's magnetic field.

Deepest Hole on Earth

The deepest artificial hole that has been made in Earth's crust, called the Kola Superdeep Borehole, is located in northwestern Russia, close to that country's border with Norway. Drilling started in 1970, with the objectives of establishing how temperature in the crust varies with depth and of finding out what rock types exist deep in the crust. It wasn't until 1989 that the hole reached its greatest depth of over 40,000 feet, where the measured temperature was higher than expected. The geologists originally planned to drill a little farther, but when they had to revise the expected temperature at a depth of nine miles upward to 570°F, this idea was abandoned, since a drill bit would no longer work at such a high temperature.

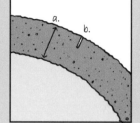

a. Baltic continental crust, 21 miles deep

b. Kola Superdeep Borehole, 7.4 miles deep

a. b.

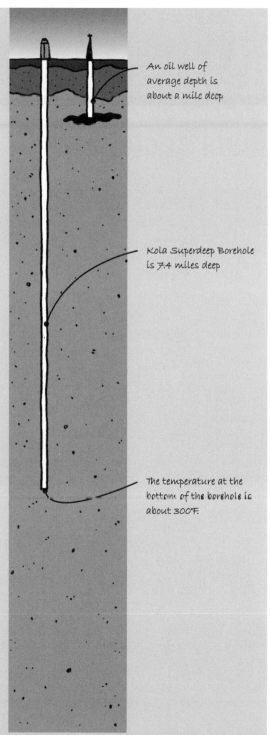

An oil well of average depth is about a mile deep

Kola Superdeep Borehole is 7.4 miles deep

The temperature at the bottom of the borehole is about 300°F.

Geothermal Energy

Earth's interior generates a colossal amount of heat, most of it coming from the decay of radioactive elements. In addition, some heat is still being dissipated from the planet's original formation. The release of this energy plays an important part in driving the movement of tectonic plates (see page 84). The heat pouring out of Earth's interior is potentially a valuable source of reliable, non-polluting energy. However, most of it emerges in just a few, mainly volcanic, regions. Some countries in regions with significant volcanic activity, such as Iceland and New Zealand, obtain a high proportion of their energy needs from this source.

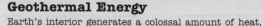

- Seven major plates make up Earth's outer shell, together with nine or ten medium-sized plates and scores of tiny "microplates."

- Plates move in relation to the underlying mantle at rates that vary from ½ inch to 3 inches a year, depending on the plate.

- Some 200 million years ago, all the continents were joined, but movements of the tectonic plates have since separated them.

- One of the largest plates—the African Plate—has started to split in two along a rift that runs down East Africa.

Plate Tectonics

Plate tectonics is the accepted theory that Earth's outermost shell, or lithosphere, consists of several plates, which fit together over its surface like a 3D jigsaw. These plates slowly move around, driven by processes occurring in Earth's interior. Many of the plates carry a continent or part of a continent. As the plates move, so do the continents. Such movements were first noticed in the early 20th century, when the process was called "continental drift," but nobody yet understood the mechanism. Eventually it was realized that huge convection currents in Earth's interior carry segments of lithosphere along at the surface in conveyor-belt fashion.

The Tectonic Conveyor

Plate tectonics works a little like a somewhat badly designed baggage conveyor system at an airport. It is made up of several individual belts, some traveling in opposite directions, others moving toward each other. The belts in the example shown below move slowly and some of the rollers are too close together, causing the belts to wrinkle and create ridges in places.

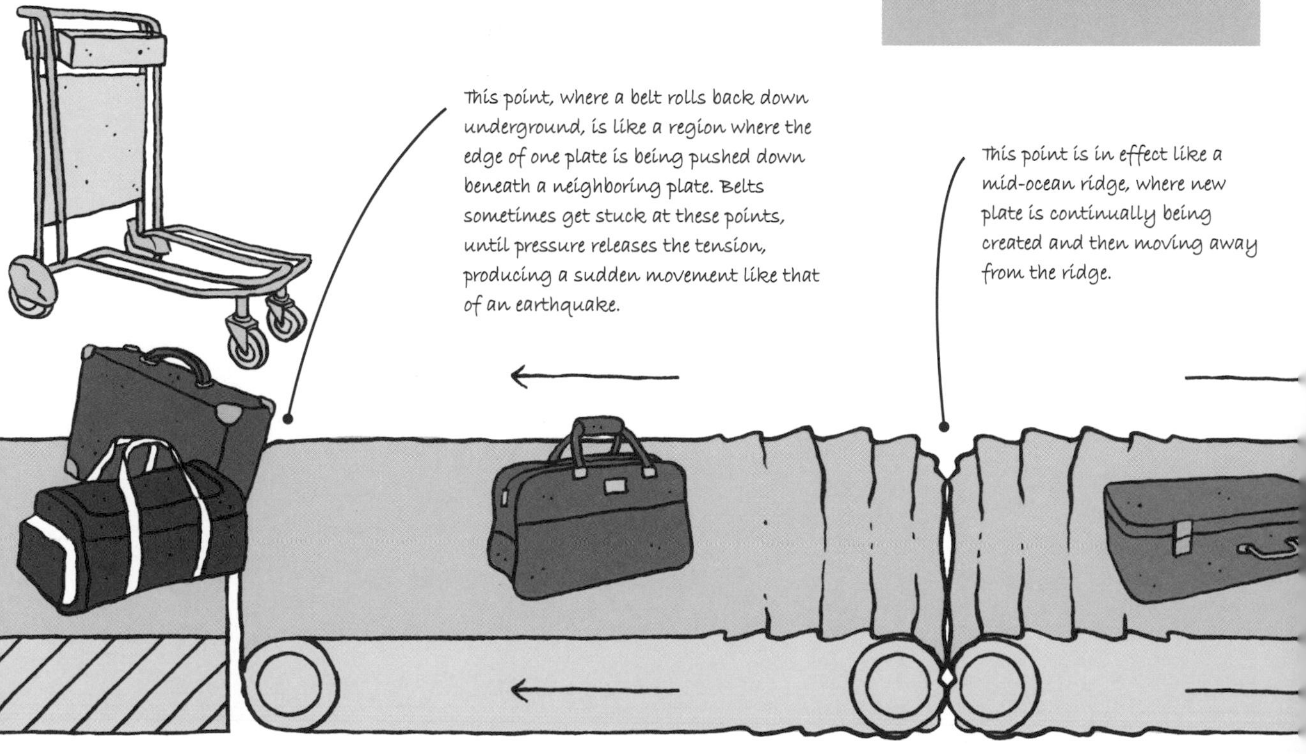

This point, where a belt rolls back down underground, is like a region where the edge of one plate is being pushed down beneath a neighboring plate. Belts sometimes get stuck at these points, until pressure releases the tension, producing a sudden movement like that of an earthquake.

This point is in effect like a mid-ocean ridge, where new plate is continually being created and then moving away from the ridge.

Earth's Plates

The plates are all different sizes and shapes. Plate movements (red arrows) can cause continents to collide so that mountain ranges such as the Himalayas are thrown up. Where one plate is pushed beneath another, deep ocean trenches form and the stresses produced cause many powerful earthquakes as well as volcanic activity. At plate boundaries where the edges of two plates are grinding past instead of colliding with each other, there are also many earthquakes.

The power supply for these rollers is like the driving force for tectonic plate movement—convective heat processes within the Earth's interior.

The effect of the convergence of two belts—a baggage pile-up—is a similar process to what happens when continents collide and a mountain range is formed.

Baggage represents continents, which are carried toward or away from each other by the moving belts.

LITHOSPHERE: Earth's rigid outer shell, which is split up into chunks called tectonic plates. It consists of Earth's crust and the uppermost layer of the mantle.

CONVECTION: A process of heat transfer that occurs in a cyclical pattern. In each cycle, warmed material rises, moves sideways, cools, and falls again.

Jargon buster

• Hundreds of small earthquakes occur around the world every day, but catastrophically destructive ones usually occur only about once a year.

• Over 95 percent of large earthquakes happen at plate boundaries where the edge of one plate pushes under the edge of a neighboring plate.

• About 50 volcanic eruptions, most of them quite small and short-lived, occur around the world each month.

• The biggest volcano on Earth, Mauna Loa in Hawaii, has an estimated volume of 18,000 cubic miles.

Earthquakes and Volcanoes

Most large earthquakes, and a high proportion of volcanic eruptions, result from processes occurring at the boundaries between tectonic plates. At their edges, plates generally don't glide past each other in a smooth manner; instead, the movement is jerky. In some areas, the opposing masses of rock are locked together by friction and sometimes do not move at all for long periods. During these periods, the forces responsible for plate movement cause a gradual build-up of strain energy (a type of potential energy) within the locked-up area. Eventually the forces and pressure become so high that a failure occurs—in other words something "gives"—and there is a sudden shift between the blocks. As this happens, energy is released in the form of powerful shock waves, producing an earthquake. In the most dramatic cases, the whole planet wobbles slightly on its axis.

Shock Wave Types

Earthquake shock waves are of various types. They include fast-moving P or primary waves, and S or secondary waves, that can travel through the solid parts of Earth's interior (P waves can also travel through the liquid parts of Earth's core). Slower-moving surface waves travel just under Earth's surface, within its crust.

Why Do Volcanoes Erupt?

As with most large earthquakes, a high proportion of volcanic activity takes place near plate boundaries, where the edge of one plate pushes under a neighboring plate. At a considerable depth beneath the surface, water escaping from the descending segment of plate lowers the melting point of surrounding rocks. These rocks turn into a hot liquid material called magma, which gradually works its way upward.

A volcano occurs when magma erupts onto Earth's surface, either fairly quietly in the form of liquid streams called lava, or more explosively in the form of clouds of ash and cinders. As well as volcanoes near plate boundaries, many others occur over what are called "hotspots." These are regions in the upper part of the mantle where narrow plumes of heat stream up from deeper regions. These heat plumes cause rocks in the area to melt, producing magma, which then erupts onto the surface.

Why Do Some Volcanoes Not Erupt?

Some volcanoes are dormant for decades or even centuries at a time, typically because a large solidified mass of lava has plugged the surface vent through which the magma had been erupting. Meanwhile, a large chamber full of magma can form and grow beneath the volcano, generating ever-increasing pressure. Eventually an event such as a small earthquake causes the lava plug to blow out, and the release of the pent-up pressure produces a dramatic eruption. Events of this kind have accounted for some of the most destructive volcanic events ever recorded. For example at Mount St. Helens, USA, in May 1980, most of the top half of a volcano was blown out, and everything over 230 square miles was burned by super-hot gas and volcanic ash.

Earthquake Prediction

Can earthquakes be predicted? The answer in terms of where an earthquake is likely to occur in the future is "yes," but predicting when an earthquake will occur is very difficult. Signs that experts look for include uplift of the ground, which can indicate strain building up in rocks. Another technique has been to look for "seismic gaps"—regions in an earthquake zone that have had little recent earthquake activity. By studying the history of earthquakes in a particular area, scientists can calculate a probability that another earthquake will occur within a given time period.

Effects of Earthquakes

The spot underground from which the energy of an earthquake emanates is called the focus, and the spot immediately above the focus, on the surface, is called the earthquake epicenter. Most of the damaging effects occur near the epicenter and derive from the violent shaking caused by arrival of the shock waves. Buildings not designed to withstand shaking are prone to collapse. Some types of soil lose their strength during a severe earthquake so that previously supported buildings sink into the ground.

Strain Build-up and Release

You can model the process that eventually leads to an earthquake by attaching a piece of elastic to a heavy brick and then pulling.

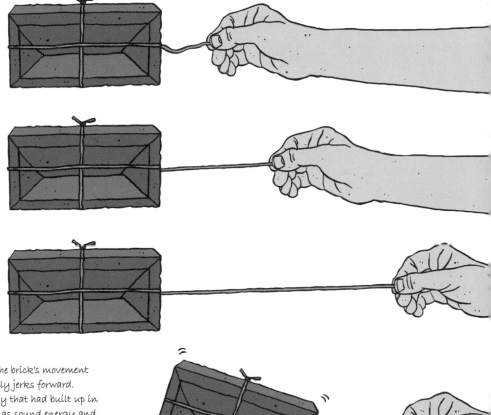

1. Place a heavy brick on the ground or on a rough-surfaced table, attach a piece of strong elastic, and pull.

2. As the elastic stretches, strain energy (a form of potential energy) builds up in the elastic. Gradually, the force tugging on the brick increases, but the brick stays still, because friction exerted by the ground (or table) opposes its movement.

3. The more the elastic is stretched, the more potential energy builds up. The pull on the brick, and the frictional force opposing its movement, also increase.

4. Finally, the friction opposing the brick's movement is overcome, and the brick suddenly jerks forward. Simultaneously, the strain energy that had built up in the elastic dissipates, reappearing as sound energy and the brick's kinetic energy. The only difference from an earthquake is that in the latter the energy converts into shock waves traveling through the ground.

bite size facts

• Over 50 separate surface currents have been identified in the oceans; some of them, such as the Gulf Stream, operate over huge distances.

• The Great Ocean Conveyor is a large-scale movement of seawater that links surface and deep-water currents. A "packet" of seawater can take centuries to complete one circuit of the conveyor.

• Ocean currents keep northern Europe 9°F warmer than it would otherwise be. In contrast, Hawaii, for example, is kept a few degrees cooler.

• Circular systems of linked surface currents, called gyres, move clockwise in the northern hemisphere and counterclockwise in the southern hemisphere.

Ocean Currents

Earth's oceans have many important effects on our lives, and not just because of the food bounty they provide. In addition to the effects of waves and tides, a series of linked currents operates throughout the oceans, transporting huge amounts of heat energy. These currents are important in modifying and stabilizing the climate in different parts of the world. They also continuously churn dissolved nutrients and other chemicals in the oceans, which is important for marine life.

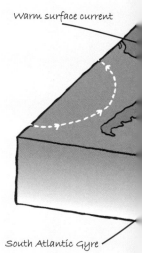

Sargasso Sea and associated garbage patch

Warm surface current

South Atlantic Gyre

The Great Pacific Garbage Patch and the Sargasso Sea

Apart from the Great Ocean Conveyor (see opposite), various wind-driven currents circulate seawater on the surface of the oceans. In the Pacific and Atlantic, in particular, several currents link up to form huge circular systems, called gyres, which rotate in a clockwise direction in the northern hemisphere and counterclockwise in the southern hemisphere. These gyres force any floating material into low-energy areas at their centers. One consequence, first noticed in the 1990s, is that a lot of plastic garbage floating in the oceans tends to become concentrated within these gyres, forming what has been called a plastic "soup." The largest of the swirling seas of plastic, covering an area bigger than

Texas, and dubbed the "Great Pacific Garbage Patch," has formed in the North Pacific. By some estimates, it contains over 100 million tons of plastic, ranging from footballs to carrier bags and toothbrushes. Another affected area is the Sargasso Sea, centered some 870 miles east of Florida, in the middle of the North Atlantic gyre. Originally famed for the weed (Sargassum) that floats on its surface, the Sargasso Sea now also harbors vast quantities of plastic.

These accumulations of floating plastic—some thrown off ships or oil rigs, with the rest originating from land—is estimated to kill more than a million seabirds every year, as well as more than 100,000 marine mammals. They accidentally consume bits of plastic, which then damage their digestive systems. The problem is likely to get worse until we become more economical in our use of plastic and more careful about its disposal.

Jargon buster

THERMOHALINE: Any type of seawater circulation in the oceans that is driven by changes in temperature or salinity (saltiness) rather than the wind.

The Great Ocean Conveyor

Part of the system of currents, known as the Great Ocean Conveyor, or thermohaline circulation, is thought to operate globally, linking relatively fast-moving, heat-laden, surface currents with deeper, slower movements of freezing cold, salty water. It has been likened to a giant conveyor belt that redistributes heat over Earth's surface. The Conveyor has a particularly important effect in the North Atlantic, through which it carries heat from the tropics. In this way, it warms northwestern Europe, giving it a much milder climate than it would otherwise have.

The main driving force for the Ocean Conveyor is the sinking of vast amounts of seawater in the extreme north Atlantic. This results from warmer water that comes up

from the south, giving up its heat in the colder waters near the North Pole and therefore becoming colder and denser. After sinking to the bottom, this water moves extremely slowly—just a few yards a day—throughout the deep oceans, eventually rising back to the surface in different parts of the Indian and Pacific oceans. From there, a series of surface currents return it to the North Atlantic. Water movements in the southern oceans link in with the system.

There are fears that global warming (see page 108) could interfere with or even shut down the operation of the Ocean Conveyor. Ironically, one of the major effects of a shutdown would be a much colder climate in northwestern Europe.

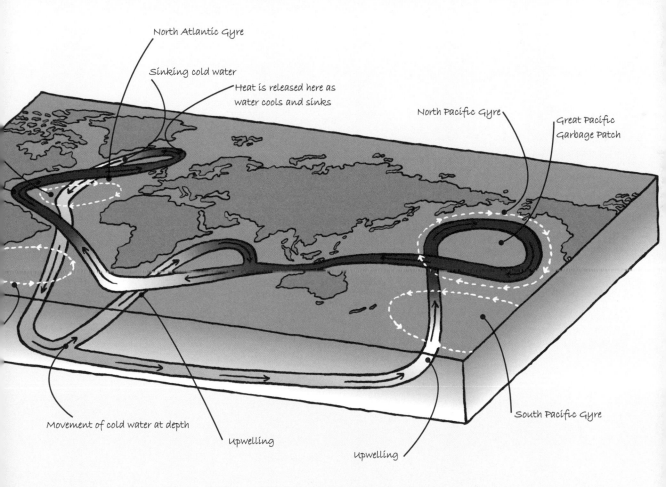

North Atlantic Gyre

Sinking cold water

Heat is released here as water cools and sinks

North Pacific Gyre

Great Pacific Garbage Patch

South Pacific Gyre

Movement of cold water at depth

Upwelling

Upwelling

Tides

Jargon buster

SPRING TIDES: Particularly noticeable and vigorous tides that occur every two weeks in which there is a large difference between high and low tide. They occur when Earth, the Moon, and the Sun are positioned more or less in a straight line.

Cyclical variations in the height of the sea relative to the land, called tides, result from gravitational interactions between the Earth, Moon, and Sun, combined with Earth's rotation. There are two basic tidal cycles: one that occurs daily, the other every two weeks. The daily cycle is the familiar one of alternating high and low tides, and results from Earth's interaction with the Moon. The fortnightly cycle is a variation in the strength of the daily tides between relatively weak "neap" tides (with a small difference between high and low tide) and more vigorous "spring" tides that follow a week later. This cycle results from Earth's gravitational interaction with the Sun working in combination with the effect of the Moon.

What Causes the Daily Tides?

The basic pattern of daily tides results from a combination of Earth's own daily rotation and its interaction with the Moon.

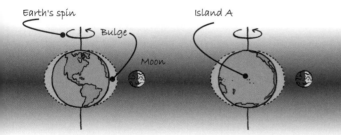
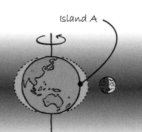

1. Earth rotates once every 24 hours around its north-south spin axis. Because of its interaction with the Moon (see panel opposite), there are slight bulges in the oceans on opposite sides of the planet.

2. As the planet spins, these bulges of water sweep over the surface. Thinking of it in another way, particular spots on the surface move into and out of the bulges. Take island A in the Pacific Ocean. This starts in a low-tide position, then...

3. ...about 6¼ hours later, it has entered one of the regions where there is a bulge in the oceans and so experiences a high tide.

4. About another 6¼ hours later, the island is in the second low tide zone around the back of the planet.

The causes of tides can be best understood by first imagining that the Earth's surface is more or less uniformly covered in water, so that the "bulges" of water in its oceans (see *What Causes the Daily Tides?* opposite), can smoothly sweep over its surface as the Earth turns. While this helps explain a lot, in reality the presence of landmasses and ocean basins of variable depth means that the actual pattern of tides in some places greatly varies from the standard pattern. This means, for example, that while many places in the oceans have two high tides a day, some have only one.

Why Is There a "Bulge" of Water on Each Side of the Planet?

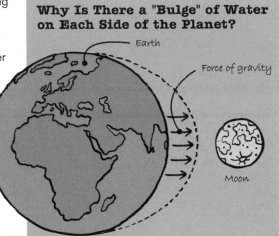

Earth

Force of gravity

Moon

Jargon buster

CENTRIFUGAL: This word means "center fleeing." For an object to move in a circle, a force must constantly be pulling that object toward the center of the circle. Acting in the opposite direction to this force, there is always what is called a *centrifugal* or *inertial* reaction, which results from the object's normal tendency to carry on moving in a straight line. The centrifugal reaction means that any fluid part of the object will move away from the center of rotation.

The cause of the bulge in the oceans on one side of Earth is easy to understand. It's the result of the Moon drawing the oceans towards it through the force of gravity (see diagram above).

The cause of the second bulge is more subtle. As the Moon orbits Earth each month, it produces a small circular "wobble" in Earth's own movement through space. During the course of this wobble, the oceans are pushed slightly away from the center of the movement by what is called a centrifugal reaction. The effect is the same as one might see from moving a bowl of soup around in a circle. The soup moves away from the center of the movement and "slurps" around the rim of the plate to the opposite side (see below).

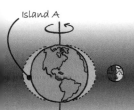

Island A

5. ...and finally, another 6¼ hours later, the island moves into the second region where there is a bulge in the oceans and so experiences another high tide. Overall, it has experienced two high and two low tides in just over a day, about 25 hours. It's not 24 hours because in the meantime the Moon has moved around the Earth a little in its own orbit.

Slurp! Slurp! Slurp!

- Global sea level is currently rising at about 3 millimeters a year.

- The current rise in sea level is being caused at least in part by global warming.

- United Nations' scientists predict that sea level is likely to rise a further 7–23 inches this century.

- A sea-level rise of around 20 inches would displace tens of millions of people from low-lying coastal areas and small islands.

Sea-Level Rise

Global sea level—a measure of the surface height of the oceans that ignores any local effects of land rising or sinking—has remained pretty stable for most of the last 3,000 years, increasing on average only a tiny fraction of an inch per year. But during the 20th century, regular measurements made using gauges positioned around coasts established that there had been an increase in the rate of sea-level rise to something between one and two millimeters per year. Since 1993, much more accurate satellite-based mapping of the shape of the oceans' surface has become available, and this has indicated that global sea level is currently rising at about three millimeters (or ⅛ of an inch) per year.

The big question now is whether global warming will cause sea-level rise to accelerate upward to dangerous levels (see the "ice in the glass" opposite). United Nations scientists predict that by the end of the 21st century, there will be a further rise in global sea level of something between 7–23 inches, depending on how much the atmosphere and oceans continue to warm. This implies that the rate of rise over the next 90 years may average between 2 and 6 millimeters per year.

Local Effects Complicate Matters

If you measure sea level against any particular land mass (by nailing a ruler to a jetty, for instance) your results can be skewed by local effects. Some continents are sinking or rising relative to the ocean around them. Areas covered in heavy glaciers during the last ice age are still rising slowly relative to the sea (the way the seat of an armchair rises after you get up).

Threatened Areas

The high end of this estimate would displace tens of millions of people living on low-lying coastal deltas. A rise of more than 40 inches would cause Bangladesh to lose one-sixth of its land area, and drown large areas of many other countries, including the Bahamas.

Worst-Case Scenario: We Lose the Ice Sheets

If global warming continues, the Greenland ice-sheet will inevitably melt completely, raising global sea level by about 23 feet. This could happen within a few centuries, and would swamp most of the world's coastal cities, including New York and London. If the entire Antarctic ice-sheet melted, it would raise sea level by a catastrophic 200 feet, but the chances of this happening in the next few hundred years are low.

When land-based glaciers and ice-sheets melt due to global warming, this adds water to the oceans...

...thus raising the water level. An even more important factor is a rise in the temperature of the oceans....

...which raises the water level further through thermal expansion.

Sea-ice cover in the Arctic and Southern oceans helps combat global warming by reflecting sunlight.

But sea-ice is diminishing, especially in the Arctic, and although sea-ice doesn't raise sea level directly when it melts...

...as it disappears, sunlight will more easily be able to warm the polar oceans and possibly accelerate the rate of sea-level rise.

> > > > Of the factors that contribute to sea-level rise, the thermal expansion of sea water is currently thought to be the most important. Contributions also come from the melting of mountain glaciers and the Greenland ice-sheet, but so far, the Antarctic ice-sheet has lost little if any of its ice overall. This situation may change as global warming continues...

• The atmosphere has a layered structure. The variations in conditions that we call "weather" occur only in the lowest layer, the *troposphere*.

• Temperature in the atmosphere varies from -130°F at a height of about 60 miles, to as much as 2,700°F at altitudes above 90 miles.

• As the Sun heats the troposphere, it causes the air in it to circulate in structures called cells. These play an important role in establishing climate zones at Earth's surface.

• Many satellites orbit in the upper layers of Earth's atmosphere, where the air is so thin that it produces almost negligible resistance to movement.

Earth's Atmosphere

The atmosphere is a cocoon of gases that surrounds our planet. Its structure, composition, and behavior affect such phenomena as climate, radio transmission, and the transfer of energy between different parts of the surface. By shielding the surface from harmful radiation, trapping heat, and providing a vast reservoir for gases such as oxygen, nitrogen, and carbon dioxide, it has played a key role in allowing life to evolve on Earth.

Atmospheric Layers

The atmosphere can be thought of as having three basic layers. The lowest, the *troposphere*, is where all weather-causing variation occurs. Above this is the relatively calm and stable (but very cold) *stratosphere*, which contains the protective ozone layer (see page 104). Above the stratosphere are the extensive upper layers of the atmosphere. Above around 50 miles from Earth's surface, much of the air has been turned into ions (electrically charged particles) as a result of excitation by ultraviolet radiation from the Sun. This region is called the *ionosphere* and it has great practical importance, since it reflects some types of radio waves. Simply bouncing the radio waves off the ionosphere allows radio messages to be transmitted much farther across Earth's surface than would otherwise be possible.

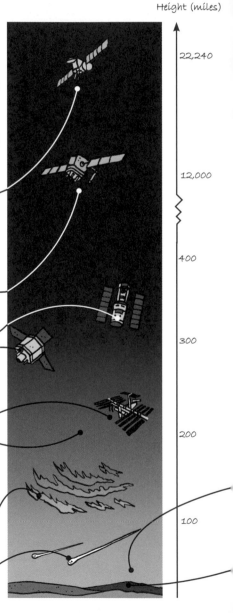

Height (miles)

22,240

12,000

Geostationary satellite. In a circular orbit above the equator, this completes one orbit every 24 hours, meaning that it constantly "hovers" over the same spot on the equator.

Uppermost layer of the atmosphere, called the exosphere, gradually merges with space.

Hubble telescope

Remote-sensing satellite in low-Earth orbit

International Space Station

The thermosphere. Part of the ionosphere, this layer consists mainly of ions (electrically charged particles).

Aurora borealis (northern lights) or australis (southern lights)

Meteor trails

400

300

200

100

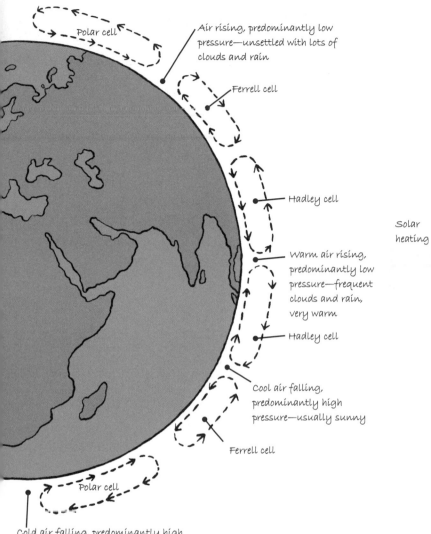

Polar cell

Air rising, predominantly low pressure—unsettled with lots of clouds and rain

Ferrell cell

Hadley cell

Warm air rising, predominantly low pressure—frequent clouds and rain, very warm

Hadley cell

Cool air falling, predominantly high pressure—usually sunny

Ferrell cell

Polar cell

Cold air falling, predominantly high pressure—sunny (summer) and cold

Solar heating

Atmospheric Circulation Cells

A good starting point for understanding how atmospheric processes affect climate and weather is to understand that heating by the Sun creates a continuous circulation of air in the troposphere.

This occurs in the "cells" shown on the left. In some parts of the world—around the equator and at latitudes of about 60° north and south of the equator—these cells create rising columns of air, associated with low atmospheric pressure at the surface and somewhat unsettled weather. Conversely, at about 30° north and south of the equator (for example, the southwestern United States), are regions where air is falling toward the surface. This causes high atmospheric pressure and a settled climate with predominantly sunny weather. The operation of these atmospheric cells is one of the basic reasons why the climate in northern Europe or northeastern North America is different from that in Egypt, Arizona, or tropical South America, for instance.

The stratosphere. Containing much thinner (lower density) air than the underlying troposphere, it is more uniform and stable. It is about 22 miles deep.

The troposphere. This is the shallow but dense bottom layer of the atmosphere. It varies in depth from five miles over the poles to 10 miles over the equator.

ATMOSPHERIC PRESSURE: Also called *barometric pressure*, this is the pressure at Earth's surface caused by the weight of the column of air above it pressing downward. At any particular spot, it varies from time to time as a result of small changes that happen continuously in the air column.

Jargon buster

Weather

The term "weather" refers to the constantly changing atmospheric conditions—in terms of factors such as temperature, wind speed, and sunshine—that prevail at a particular place over a period of time. Changes in weather can be rapid—as when a wet morning turns into a sunny afternoon—but over a long period, the average weather conditions are less changeable. The average weather in a particular place over many years makes up the climate for that location.

Most weather, particularly in temperate regions of the world, is explainable from the behavior of localized centers of high and low pressure that move across the surface. High-pressure cells tend to be slow-moving and associated with fine weather. Depressions (low-pressure cells) are faster-moving and frequently linked to clouds and rain. The latter is primarily because low pressure results from rising warm air, and if this air is moist at all, any water vapor in it will turn into liquid droplets as it cools, thus forming clouds and ultimately rain.

Closely linked to high- and low- pressure cells are different masses of air, some warm and some cool, that also move around on the surface. These frequently chase each other around centers of low pressure, driven by surface winds. The boundaries between different air masses are called "fronts" and the passage of different types of front over the ground produces some characteristic weather patterns.

1. Previously on Planet Earth...

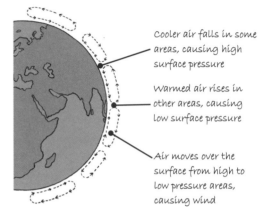

Cooler air falls in some areas, causing high surface pressure

Warmed air rises in other areas, causing low surface pressure

Air moves over the surface from high to low pressure areas, causing wind

Jargon buster

TEMPERATE REGION: Parts of the world between the tropics and polar regions (the Arctic and Antarctic).

Cold Front

The approach of a cold front brings a steady drop in atmospheric pressure and billowing clouds, followed by gusty winds and thunderstorms (sometimes severe), with heavy rain, hail, or snow, as the front passes through. This is followed by a drop in temperature and a rise in atmospheric pressure.

Warm Front

The approach of a warm front brings a gradual lowering of cloud cover and intermittent rain or snow. This is followed by clearing of the cloud cover and a rise in temperature as the warm air mass arrives.

2. Viewed differently...

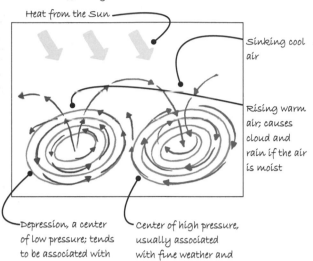

Heat from the Sun

Sinking cool air

Rising warm air; causes cloud and rain if the air is moist

Depression, a center of low pressure; tends to be associated with unsettled weather and cloud

Center of high pressure, usually associated with fine weather and clear skies

3. How it may look on a map.

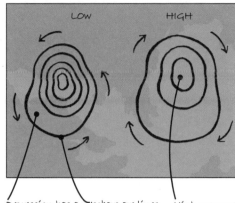

LOW   HIGH

Depression has a counterclockwise wind pattern around it (in northern hemisphere)

Isobars are lines plotted on a map that connect places where the atmospheric pressure is the same at any one moment

High-pressure cell has a clockwise wind pattern around it (in northern hemisphere)

4. Associated with depressions and high-pressure cells are patches of warmer and cooler air, which tend to chase each other around the centers of low pressure, pushed by winds.

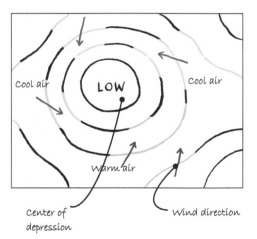

Cool air   LOW   Cool air

Warm air

Center of depression

Wind direction

5. On maps, the boundaries between the masses of cold air and warm air are called "fronts." The approach and passing of each type of front brings characteristic patterns of weather, lasting from a few hours to a few days.

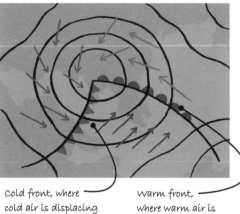

Cold front, where cold air is displacing warm air

Warm front, where warm air is displacing cold air

MAP SYMBOLS	Warm front	●●●
	Cold front	▲▲▲

- A lightning bolt travels through air at 80,000 mph.

- The electrical current in a lightning bolt can be as much as 200,000 amps.

- The electrical charge or potential difference within a thundercloud that generates lightning can be as much as 100 million volts.

- Cloud-to-ground lightning heats the surrounding air to more than 54,000°F, five times hotter than the surface of the Sun.

- In some cases, a lightning flash involves up to 20 lightning bolts between cloud and ground.

Lightning

Nearly all forms of lightning originate in large thunderclouds, which act like giant electrical generators. These clouds are affected by strong, turbulent internal winds, which hurl around tiny ice crystals, smashing them together in countless collisions that create static electricity. This turns the thundercloud into something like a huge battery with positive charges at the top and negative charges at the bottom.

STATIC ELECTRICITY: A buildup of electric charge on an object, usually caused by friction or collisions that remove electrons from one object and deposit them on another.

Jargon buster

How does Cloud-to-Ground Lightning Happen?

1. In a thundercloud, partially frozen water collides with hard ice crystals. This causes electrons to transfer to the denser partially frozen water, which falls to the base of the cloud. The hard ice crystals become positively charged and the icy water negatively charged. As it passes overhead, the cloud induces a positive charge on the ground.

2. Electrons leak down from the cloud toward the ground, partially ionizing (causing electrical charge separation) in the air. The pencil-thin ionized channel that they move down is called a step leader. Simultaneously, positive charges climb up tall objects on the ground, such as trees or towers, to meet the step leader.

How Far Away?

The flash from a lightning bolt travels at the speed of light, so we see it almost instantly, whereas the sound of thunder travels at about 1,100 feet per second. This means you can calculate how far away a thunderstorm is, in miles, by counting the seconds between the lightning flash and the roll of thunder and then dividing by five.

If You Are Caught Outside During a Storm...

Get inside a building or car and close the windows and doors. Cars provide excellent protection because their metal shells pass current around the outside. If there is no shelter, stay away from isolated trees, because they are natural lightning rods. Also avoid standing above the surrounding landscape, such as on an exposed hilltop, or being out in a small boat.

Can Lightning Strikes Cause Planes to Crash?

On average, lightning strikes a commercial plane twice a year, but aircraft are generally quite safe. Although passengers might hear a bang, see a flash, and feel an alarming jolt, the metal exterior of the plane conducts the current over the outside without harm. Nonetheless, aircraft can be vulnerable to lightning. Instruments inside the cockpit are sensitive to the massive surge of electromagnetic energy associated with a lightning strike, and these instruments may then give false readings. Extremely rarely, a lightning strike can puncture the metal exterior of an aircraft and even ignite the fuel.

Types of Lightning

Most commonly, lightning occurs within a cloud and involves a discharge between oppositely charged electrical regions. This is seen as a flickering glow. Lightning can also occur between clouds. But the best understood, most spectacular type of lightning is the cloud-to-ground variety (see below).

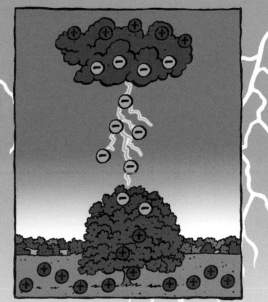

3. When the positive charges meet the step leader, a powerful electrical discharge—a bolt of lightning—flows through the ionized channel. During the discharge, there is a massive downward movement of electrons. The electrons nearest the ground descend first, then the electrons a little higher up, and so on.

4. The discharge drains electrons from the base of the cloud. Additional lightning bolts may occur in rapid succession along the same channel while there is still a significant charge difference between the cloud and ground. Later, further strikes can occur because of the positive charges isolated at the top of the cloud. These lightning bolts are called positive lightning.

6

Environment Stuff

CHAPTER CONTENTS

Concerns about damage to the environment and what to do about it are among the most discussed scientific issues of our age. Environmental concerns tend to go through phases. Back in the 1960s, for example, there was particular disquiet about the effects of insecticides on birds and other wildlife. In the 1970s, anxieties centered around acid rain and water pollution, while in the 1980s and 1990s, attention switched to issues such as thinning of the ozone layer, concerns about the disposal of nuclear waste, deforestation, and loss of biodiversity.

The Environment Today

None of these problems have gone away, and some have gotten worse. In a few cases, such as ozone thinning, the problem has been reduced through concerted international action. But in recent years, most of the above concerns have been dwarfed by preoccupation with the major environmental issue of the 21st century—global warming and the various undesirable effects that it is predicted to have, such as climate change and sea-level rise.

An Energy Issue

Global warming is essentially an issue to do with energy—how much we use, and the sources from which we obtain it. For over a century, we have run our industries, heated our homes, and powered our means of transport with energy obtained mostly from fossil fuels, notably coal, oil, and gas. But burning these fuels emits carbon dioxide, a "greenhouse gas" whose increasing concentration in the atmosphere is acknowledged by the vast majority of scientists to be the main cause of warming. Reflecting these current concerns, much of this chapter is devoted to technologies that provide a hope of combating global warming. This includes discussion of renewable energy sources, such as wind and solar power, the use of biofuels, and electric vehicles, which provide some prospects of reducing our dependence on fossil fuels. This is not to forget that numerous other environmental issues also exist, and so parts of this chapter have been devoted to such topics as ozone depletion and what is being done about the radioactive waste generated by nuclear power plants.

• The main sources of human-caused air pollution are power stations, vehicle emissions, and agricultural burning.

• The World Health Organization estimates that about two million premature deaths are caused each year by air pollution in cities across the world.

• Polluted air is particularly damaging to those who have lung and heart disease.

• The removal of lead from gasoline in the 1990s has been a notable success in the fight against air pollution.

Air Pollution

Carbon Dioxide (CO_2)

A colorless gas, carbon dioxide is considered a pollutant because it is a "greenhouse gas" and the main contributor to global warming (see pages 108–111). The main sources of carbon dioxide in the atmosphere include natural events such as volcanoes and human activity—notably the burning of fossil fuels such as gasoline and coal.

Particulates

Particulates are fine particles floating in the air. Some occur naturally, originating from forest fires, volcanic eruptions, or dust storms. Others result from the burning of fossil fuels in power plants or are produced in car exhausts. A high level of particulates increases the risk of lung disease. Some of the most polluted cities have concentrations up to six times higher than healthy levels.

Carbon Monoxide (CO)

A colorless, odorless gas, carbon monoxide is produced by wood fires and is also present in cigarette smoke. The gas is toxic because it reduces the blood's ability to carry oxygen around the body. Most carbon monoxide in city air comes from car fumes. In some cities the gas is present at a level of 150 parts per million (good air quality requires a level below 10 ppm).

Ozone (O_3)

Ozone is an uncommon form of oxygen that has three atoms in each of its molecules. Ozone in the stratosphere protects life on Earth by absorbing excess solar radiation (see pages 104–105). Near the ground, it can harm the lungs. In the lower atmosphere, ozone contributes to global warming.

Air pollution is the release into Earth's atmosphere of substances that can cause harm or discomfort to living things. Pollutants are most often gases, but they can also take the form of fine solid particles (as in smoke) or tiny liquid droplets. Some air pollution results from natural phenomena such as volcanic eruptions, but a more important factor is human activity, including coal-burning power stations, vehicle exhausts, and agricultural burning.

The seven air pollutants highlighted here are among the most important. Others include ammonia, volatile organic compounds (VOCs), and some toxic metals. A particularly noxious form of air pollution, photochemical smog, tends to develop in places that have both a dry, sunny climate and a huge number of automobiles—for example, Mexico City and Los Angeles.

Nitrogen Dioxide (NO_2)

This acrid gas produced by power stations is present in vehicle exhausts and plays a part in ozone formation. It contributes to the formation of acid rain. Good air quality specifies a concentration below 30 parts per billion. Higher levels can damage the lungs.

Sulfur Dioxide (SO_2)

A colorless, pungent gas produced by many industrial processes, sulfur dioxide irritates the lungs. In the air, it combines with tiny water droplets to form dilute sulfuric acid, a major component of "acid rain," which is harmful to plants and aquatic animals. In some cities, levels reach over 200 parts per billion, whereas for good air quality, a level below 50 ppb is required.

Lead

Once a major pollutant in car exhaust fumes, emissions of lead from this source have dropped by 98 percent since leaded gasoline was phased out. It is still emitted by some metal processing and battery processing plants. If breathed in, lead can damage many parts of the body and slows mental development in children. A concentration below 0.2 parts per billion is desirable.

- Ozone forms a protective layer in the stratosphere that prevents most harmful UV-radiation from reaching Earth's surface.

- The protective ozone layer has been thinned or depleted over the past 50 years or so by atmospheric pollutants, principally chemicals called CFCs.

- The total amount of ozone in the stratosphere has always been small—if it was all collected together in a layer of pure ozone, this would be just a few millimeters thick.

- The Antarctic ozone hole is a seasonal reduction of up to 70 percent in the amount of ozone over Antarctica.

Ozone Depletion

Some 12–16 miles above Earth's surface, in the layer of the atmosphere called the stratosphere, is a concentration of the unusual form of oxygen called ozone, or O_3.

The Culprits

The main ozone-depleting pollutants were identified in the 1970s as chemicals called chlorofluoro-carbons (CFCs). These are a group of chemically similar gases used in aerosols, refrigeration systems, air conditioners, and solvents. The chemical reactions that destroy ozone are complicated, but typically involve the release of free chlorine atoms from CFCs; these free chlorine atoms then act as catalysts for the breakdown of ozone. It takes only a small amount of a CFC to destroy a huge amount of ozone.

Unfortunately, CFCs were being used and released into the atmosphere for many years before the full extent of the harm they were doing was realized. Even if their production and release is now reduced to zero, it will be decades before they completely disappear from the atmosphere. Potential problems still remain today from the huge quantity of refrigeration and air conditioning systems in the world that still contain CFCs. These need to be disposed of correctly to prevent a setback in the progress that has been made so far in combating ozone depletion.

Ozone is continually being formed at this altitude by chemical reactions between normal oxygen molecules (O_2). Soon after they form, ozone molecules absorb some energy in the form of short-wavelength ultraviolet (UV) radiation from the Sun, and later break down again to ordinary O_2 and emit the stored energy as heat. In this way the ozone acts as a sort of screen, preventing most short-wavelength UV radiation from reaching Earth's surface. This type of UV radiation can damage plants and cause skin cancer in humans—therefore, any depletion of the ozone layer is potentially dangerous.

During the 20th century, the ozone layer became thinner due to a buildup of various atmospheric pollutants that destroy ozone. By 1985 a serious seasonal degree of thinning over Antarctica,

Antarctic ice-sheet .

Why the Antarctic?

Ozone depletion occurs most strongly over Antarctica due to the unique weather conditions that develop there during the Antarctic winter. The air over Antarctica becomes exceptionally cold—a state of affairs that is maintained by a circular system of winds around the edge of the continent (the "polar vortex") that isolates the cold air mass.

These conditions allow the development of special clouds called polar stratospheric clouds (or PSCs) in the stratosphere. These contain ice particles, the surfaces of which provide a site for reactions that break down ozone. By the time spring arrives and the Sun returns after the polar night, the ozone levels have been reduced.

Antarctic ice-sheet

"Hole" in ozone layer

Ozone layer

known as the "ozone hole," was being reported. This "hole" still develops each year, with its greatest extent occurring in September and October (spring in the southern hemisphere). A less pronounced degree of ozone depletion occurs in winter and spring each year over the Arctic.

In response to ozone thinning, an international treaty called the Montreal Protocol was signed in 1987 by most of the world's nations, agreeing to drastic reductions in emissions of ozone-depleting pollutants. Encouragingly, as a result of this action, the damage is now gradually being reversed, and current data suggest that the ozone layer might be back to normal by the year 2050, as long as the production and release of the damaging pollutants remains controlled.

Protocol Revisions

The Montreal Protocol on the phase-out of ozone-depleting substances, and their replacement with acceptable substitutes, has been revised seven times since 1987. Because of its widespread adoption, it is regarded as one of the most successful international agreements ever implemented.

CATALYST: A substance that initiates or speeds up a chemical reaction without itself being changed by that reaction.

Jargon buster

The ozone hole was first discovered by scientists working at a British Antarctic survey station in the early 1980s.

- Biomagnification is the buildup of a toxic substance in a food chain in which the toxin becomes increasingly concentrated with each stage in the chain.

- Most large marine animals contain high concentrations of toxins in their tissues due to the effects of biomagnification.

- Pregnant women and children are advised not to eat some species of large fish, such as shark and swordfish, because of their exceptionally high mercury content.

- The numbers of some predatory birds—for example, bald eagles—have recently recovered, due to bans on toxic substances that can biomagnify.

Biomagnification

An insidious process that particularly affects lake and marine environments (although a similar process can also occur on land) is that of *biomagnification*. This is the gradual buildup of a toxic substance from seemingly innocuous to potentially harmful concentrations as a result of the operation of a food chain. The animals most affected by this process are those at the top of food chains—particularly (in the case of marine pollution) seabirds and marine mammals such as dolphins, seals, and porpoises. Biomagnification is also potentially a cause of harm to humans through the consumption of affected fish.

A number of specific toxic chemicals are particularly likely to be biomagnified if they enter the environment because of their persistence (i.e. they do not break down easily) and because, once they get into animals, they are neither broken down nor excreted, but instead become concentrated in the animal's fatty tissues. These chemicals include some pesticides (such as DDT), compounds of mercury and arsenic, and a class of compounds called polychlorinated biphenyls (PCBs), which were once a common component of various industrial and household chemicals.

Seabirds Suffer Too

Like marine mammals, seabirds are at the top of marine food chains, so they are particularly susceptible to biomagnifying toxins. When these build up in the bodies of seabirds, they can reduce fertility and also interfere with the process of eggshell formation. This causes eggs to be laid that have thinner shells than normal, and can lead to the birds crushing their eggs instead of incubating them. This is known to have caused major population declines in some species, including pelicans and petrels, though this problem has declined somewhat since the 1970s due to reduced use of certain pesticides, notably DDT.

1. The concentration of a toxic pollutant in sea water might seem to be exceedingly low, perhaps five parts per trillion by weight.

2. Phytoplankton (small plants that float in the ocean) absorb the toxin. It remains in their tissues, without being broken down or excreted. Over time the toxin builds up to a concentration of for example, 200 parts per trillion—a 40-fold increase.

3. Zooplankton (small, floating marine animals) consume the phytoplankton and so take in the toxin. Again, it remains locked in their tissues, without being broken down or excreted. Its concentration builds up to perhaps two parts per billion, a ten-fold increase over the concentration at the previous stage.

FOOD CHAIN:
A sequence of groups of living organisms, each group being a source of food for the next group in the sequence. The final group, which is not eaten by anything else, is said to be at the "top of the food chain."

Jargon buster

4. Small fish graze on the zooplankton, and so take in the toxin, which becomes concentrated in their fatty tissues. There it builds up to a concentration of around 20 parts per billion, another ten-fold increase.

5. Larger fish consume the smaller fish, and again take in the toxin, which becomes concentrated to 80 to 100 parts per billion in their fatty tissues, a four to five-fold increase.

6. A dolphin eats the larger fish. The toxin gradually becomes concentrated in tissues such as its liver to harmful levels of 10,000–15,000 parts per billion. This may reduce the animal's fertility and make it more susceptible to disease.

Toxic pollutant

bite size facts

- Earth's surface temperature has increased by approximately 1.5°F over the past century.

- The vast majority of climate scientists believe this warming is the result of an increase in substances called greenhouses gases in the atmosphere, caused at least in part by human activity.

- The level of atmospheric carbon dioxide (CO_2), the most important greenhouse gas, has increased by over 20 percent since 1960.

- Based on different estimates of future greenhouse gas emissions, scientists predict a further increase in average global surface temperature of 2–12°F by the end of the 21st century.

Global Warming

The familiar term "global warming" refers to an increase that has occurred in the average temperature of Earth's atmosphere and the surface of its oceans since the early 20th century—in other words, over the past 100 years or so. The evidence that this warming is happening is strong and includes temperature records showing a clear temperature rise over the past 100 years; a pattern of glacier melting and measured sea-level rise that corresponds with predictions about the warming; and an increase in the number of extreme weather events, which also agrees with computer predictions of the effects of warming.

Most climate scientists are convinced that the cause of this warming is a rise in the level of substances called greenhouse gases in the atmosphere, particularly carbon dioxide (CO_2) (see also pages 102–103). The phenomenon, known as the "greenhouse effect," explains how gases such as CO_2 and methane, which normally keep the atmosphere warm, can cause overheating if their levels in the atmosphere rise too high. Careful measurements of the level of atmospheric CO_2 shows that it has indeed been rising at a steady rate since at least 1958.

Consequences

Some of the serious consequences that scientists predict from continued global warming include:

- A further rise in sea level (see page 92).

- An increase in the number and severity of extreme weather events.

- In some areas, more frequent droughts with follow-on effects on agricultural yields and an increased risk of famine.

- Acidification of the oceans, due to their absorbing more carbon dioxide from the atmosphere, and the consequent widespread destruction of coral reefs.

- The spread of insect-borne disease from the tropics to previously unaffected areas.

World Temperature Graph

This graph shows how global surface temperature has varied since 1850. It is clear that there has been a gradual and persistent rise since around the year 1910, with only a slight downward "blip" in the 1940s.

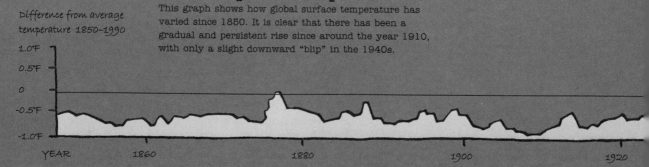

Difference from average temperature 1850–1990

1.0°F
0.5°F
0
-0.5°F
-1.0°F

YEAR   1860   1880   1900   1920

The Greenhouse Effect

The greenhouse effect is caused by a layer of gases in the atmosphere that trap heat radiated from the Earth's surface and "re-radiate" it back, thus warming the lower atmosphere. This layer of gases works in the same way as the glass in a greenhouse.

3. With an optimum level of greenhouse gases in the atmosphere, some of the longwave radiation re-radiated by Earth's surface escapes...

Normal level of greenhouse gases

Excess level of greenhouse gases

1. Incoming solar radiation (mainly short wavelength) penetrates the layer of greenhouse gases and heats Earth's surface.

4. ...and some is reflected back toward Earth's surface.

2. Longwave radiation (heat) is re-radiated from Earth's surface.

5. But with an excess of greenhouse gases, too much heat is returned toward Earth's surface and not enough escapes.

SHORTWAVE AND LONGWAVE RADIATION: Shortwave radiation, in the context of Earth's energy balance, is light and near-visible infrared and ultraviolet radiation from the Sun. Longwave radiation is lower-energy infrared (heat) radiation that leaves Earth.

Jargon buster

1940   1960   1980   2000

bite
size
facts

• The main greenhouse gases that human activity has added to the atmosphere are carbon dioxide, methane, and nitrous oxide.

• Burning of fossil fuels, deforestation, agricultural activity, and emissions from landfill sites are the main sources of these gas emissions.

• The most abundant greenhouse gas of all is actually water vapor, but human activity has no significant effect on its average level in the atmosphere.

• As well as being a greenhouse gas, nitrous oxide (also known as "laughing gas") is an important cause of ozone depletion (see page 104).

Sources of Greenhouse Gases

The Three Main Greenhouse Gases

Carbon dioxide (CO_2) makes up about 74 percent of the greenhouse gas emissions that are caused by human activity, methane about 16 percent, and nitrous oxide about 9 percent. The chart on the right shows the relative contributions of different human activities to emissions of these three greenhouse gases.

Main Sources of Greenhouse Gas Emissions from Human Activity

Carbon Dioxide Emissions

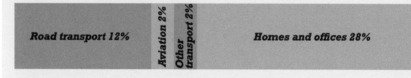

| Road transport 12% | Aviation 2% | Other transport 2% | Homes and offices 28% |

Nitrous Oxide Emissions

| Industry 6% | Deforestation and other changes in land use 26% |

Methane Emissions

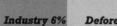

| Homes and offices 5% | Deforestation and other changes in land use 7% | Fossil fuel production 30% |

The figures (above) for industry and for homes and offices include emissions that result from the generation of electrical power for each.

Jargon buster

SEQUESTERING:
This refers to the process of "locking up" a gas (mainly CO_2) within solid material through biological or physical processes.

Where have the extra carbon dioxide (CO_2) and other greenhouse gases that have entered the atmosphere over the past 100 years come from? Natural emissions of CO_2, such as those produced by volcanic eruptions, are undoubtedly greater than those produced by human activity, but over periods longer than a few years CO_2 emissions from these natural sources have been closely balanced by the absorption of the gas by natural "sinks" such as the oceans and green plants. Climate scientists are therefore confident that most of the measured CO_2 rise has come from the burning of fossil fuels—coal, crude oil, and

Naysayer Arguments

Today, nobody seriously disputes that global warming is happening, nor that there has been a rise in CO_2 and other greenhouse gases in the atmosphere since the 1950s, which is at least partly the result of human activity. What is disputed by some skeptics, however, is the assumption that the CO_2 rise is causing the warming. They argue that some other factor unconnected to human activity might be causing the warming and that at least part of the rise in CO_2 results from (rather than is a cause of) the warming—perhaps through oceans releasing CO_2 as they warm. If this were the case, there may be nothing much people can do to stop or slow the

warming. These "naysayer" arguments are rejected by the vast majority of climate scientists, however. One reason is that chemical studies of atmospheric CO_2, based on the study of isotopes (see page 16), show that the CO_2 increases are coming from the burning of fossil fuels and not from the oceans (which are actually taking up CO_2, not releasing it). Furthermore, no "naysayer" has yet to come up with a convincing explanation of what might be causing the warming other than the increase in greenhouse gases in the atmosphere. Proposed alternative causes, such as increased solar activity, are not supported by the available data.

Industry 31%

Deforestation and other changes in land use 17%

Fossil fuel production, mining, and distribution 8%

Agricultural activities and by-products 62%

Waste disposal and treatment 2%   Other 4%

Agricultural activities and by-products 40%

Waste disposal and treatment 18%

natural gas—used to provide energy for transport and to generate electrical power and heating for industry, homes, and offices. Much of the rest of the extra CO_2 in the atmosphere has come from deforestation—the clearing of forests and the burning of wood. This releases carbon that was previously "locked up" in trees in the form of cellulose or other carbon-containing chemicals. Some additional CO_2 emissions come from the mining, production, and distribution of fossil fuels.

Methane Sources

Other than CO_2, the most important greenhouse gas added to the atmosphere by human activity is methane (CH_4). This is an even more potent greenhouse gas than CO_2. The main sources of methane emissions related to human activity include livestock farming (cows emit methane), landfill sites (rotting biological materials release the gas), and mining activities (coal seams always contain some methane, which is released by mining).

- One hour of solar radiation falling on Earth, if properly harnessed, should in theory be capable of meeting world energy demands for a whole year.

- An average-sized wind turbine can produce enough electricity to provide power for several hundred homes.

- Nearly all energy needs in Iceland are supplied from renewable sources— mainly from geothermal energy supplemented by hydroelectric power.

- The European Union target is to supply 20 percent of its energy from renewables by 2020. The UK and USA each have a target of 15 percent.

Renewable Energy

Wind Power

Although wind power provides less than one percent of global energy needs (as of 2009), it is currently one of the fastest growing renewable energy sources, with global installed wind power capacity increasing at a rate of over 15 percent annually. In many respects, wind power typifies both the benefits and some of the short-term drawbacks of renewables. As a power source, wind energy is highly attractive because it is plentiful, widespread, clean, and produces no greenhouse gas emissions. Against that, the construction of wind farms is not welcomed everywhere due to their visual impact. In addition, the expense of installing them is high: a wind farm with a capacity of about 100 megawatts typically costs hundreds of millions of dollars to set up—more if offshore— although the costs are coming down. The energy provided by wind farms is also intermittent. However, the period of peak demand (the winter months) generally matches the period of peak supply—unlike solar power, for example.

Renewable energy is energy generated from resources that are naturally replenished, such as sunlight, wind, rain, the heat from Earth's interior (geothermal energy), and ocean waves and tides. It also includes energy derived from recently dead plant material or biomass (see pages 116–117), but excludes fossil fuels such as coal, oil, and natural gas.

The main advantages of renewables over fossil fuels are first, that they generate far less carbon dioxide (CO_2), and so are a potentially valuable means of combating global warming. Secondly, they won't run out—most are derived (directly or indirectly) from radiant energy provided by the Sun, which is going to shine for several billion more years. If properly harnessed and developed, renewables should eventually be capable of completely replacing fossil fuels.

There are some downsides to renewables, mostly short-term. In particular, renewable energy projects can be expensive to set up, and it can be many years before they achieve net reductions in CO_2 emissions. Some forms provide only intermittent power, although this problem can be solved by intelligently linking the output of different renewable power sources into a grid.

MEGAWATT: A rate of energy production of a million watts, enough to power a few hundred homes in an average developed country. A thousand megawatts is called a gigawatt and a million is called a terawatt. The world's rate of energy consumption is currently (in 2009) running at about 17 terawatts.

jargon buster

How the World Currently Supplies its Energy Needs

Currently, renewables supply only about 16.5 percent (about one sixth) of the world's energy needs, although that figure is gradually growing. They fill a little more than the two lowest rows of the chart below. Fossil fuels provide about 78.5 percent (nearly four fifths) of the energy needs, and the remaining five percent is supplied by nuclear power.

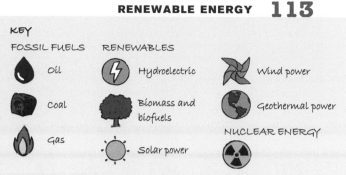

KEY

FOSSIL FUELS

Oil

Coal

Gas

RENEWABLES

Hydroelectric

Biomass and biofuels

Solar power

Wind power

Geothermal power

NUCLEAR ENERGY

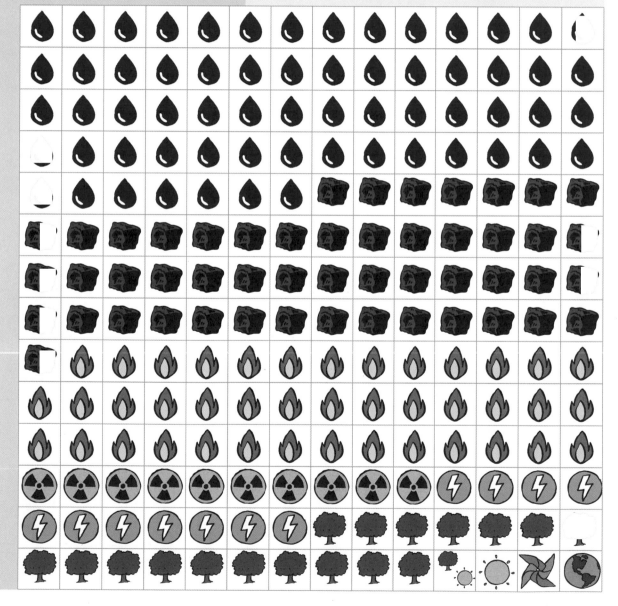

- Nuclear waste is radioactive material that largely comes from nuclear reactors or is a by-product of nuclear weapons production.

- Some 20 countries currently hold about 125,000 tons of the most dangerous waste in temporary storage, while they decide what to do with it.

- Some components of radioactive waste will still be highly radioactive in a million years' time.

- It is believed that most of the dangerous radioactive waste will eventually be buried deep underground.

Nuclear Energy Waste

Other Methods

Some scientists think that in future the problem of disposing of HLRW might be reduced through a procedure called *transmutation*. This would involve using additional nuclear reactions to process the waste and reduce its volume or hazard level. But implementation of this form of processing looks to be several decades away. Sending HLRW into the Sun could be an attractive method of disposal but is not possible at present because of the unacceptably high risk of an explosion during the rocket launch.

One of the major environmental concerns of our age is how the radioactive waste from nuclear power plants and weapons programs built up in various countries over several decades is to be disposed of. The most hazardous material, called high-level radioactive waste (HLRW), consists mainly of spent fuel from nuclear reactor cores or material derived from this spent fuel. It is intensely radioactive and dangerous to living things. It also generates a lot of heat.

The Scale of the Problem

The different radioactive isotopes in HLRW have different half-lives (see page 18), which means they remain hazardous for different periods of time. Some have half-lives of up to a million years or more, so any

What Happens to the Most Dangerous Waste?

Although the diagram on the right gives an overview of the process, the exact details vary between countries. So far, no country has begun the final stages of geological disposal. Many are still trying to decide on suitable disposal sites. Some HLRW has, however, reached the stage of being vitrified (dispersed in glass).

Hot, intensely radioactive spent fuel rods from nuclear reactors

Fuel rods are stored in racks in a large spent fuel water pool for up to 20 years while they cool down and lose some radioactivity.

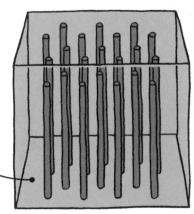

In some countries, spent fuel is then reprocessed in a reprocessing plant; this creates more liquid HLRW.

long-term strategy for dealing with HLRW has to have a timeframe of hundreds of thousands to millions of years. While most countries holding HLRW expect eventually to store it in locations deep underground, only a few—Finland, Sweden, and France among them—have made a definite decision on where they will store it and have started to implement their plan. Others, despite years of studying the problem, have yet to decide where they will store their nuclear waste. This is connected to the difficulty in finding sites where scientists can be 100 percent certain that the waste will not be disturbed for a million years.

Environmental Concerns

The general plan in most countries for dealing with HLRW is outlined below. There are environmental worries about several stages in the implementation of this process, especially about possible accidents or terrorist attacks at spent fuel pools, where a failure of cooling systems could lead to a fire and major release of radioactivity. There are also concerns about accidents during the transport of HLRW, and worries about the safety of long-term storage in sites underground.

REPROCESSING: Chemical treatment of spent fuel rods from nuclear reactors to recover still-usable plutonium and uranium.

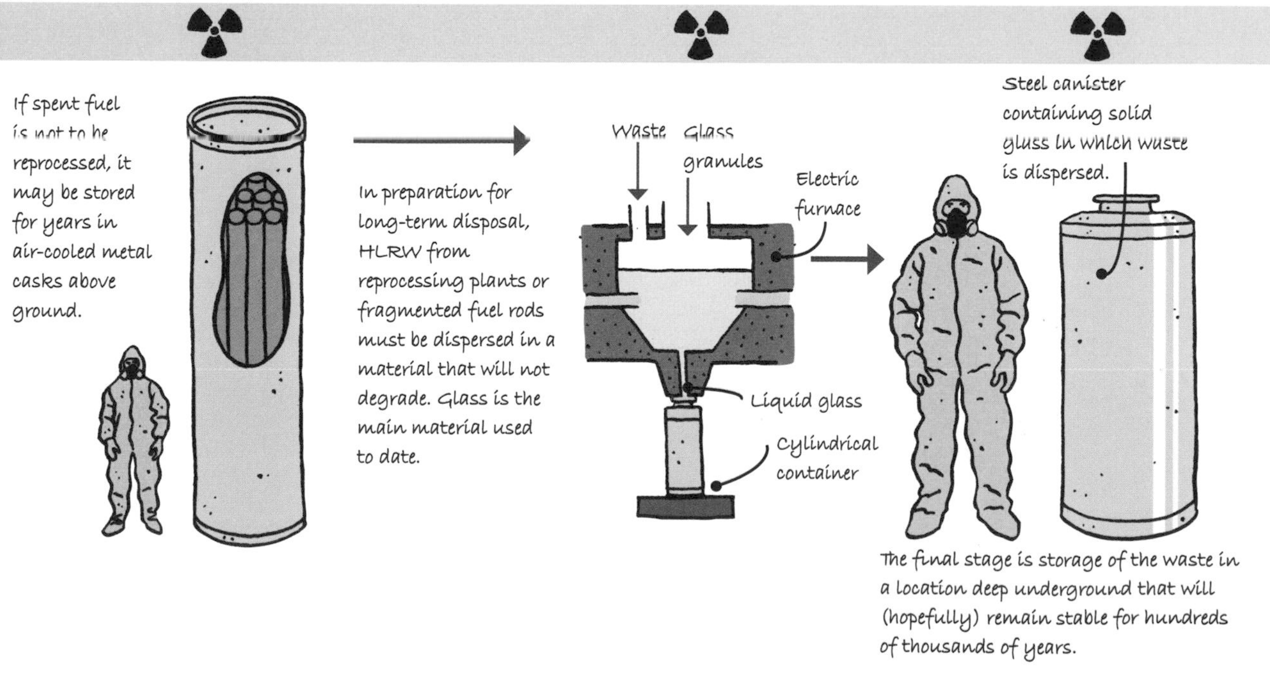

Jargon buster

If spent fuel is not to be reprocessed, it may be stored for years in air-cooled metal casks above ground.

In preparation for long-term disposal, HLRW from reprocessing plants or fragmented fuel rods must be dispersed in a material that will not degrade. Glass is the main material used to date.

Waste   Glass granules

Electric furnace

Liquid glass

Cylindrical container

Steel canister containing solid glass in which waste is dispersed.

The final stage is storage of the waste in a location deep underground that will (hopefully) remain stable for hundreds of thousands of years.

• Biofuels can be used as vehicle fuels or combusted to heat buildings or to produce electricity in power plants.

• Use of biofuels for road transport can produce an overall saving in CO_2 emissions of anything between 20 and 90 percent.

• World production of bioethanol, a common biofuel, is now running at over 17 billion gallons a year and increasing at a rate of over 30 percent a year.

• A crop area the size of the US would have to be planted with biofuel crops to power all the world's cars.

Biofuels

Biofuels are solid, liquid, or gaseous fuels that are derived from recently living biological material (biomass), as opposed to fossil fuels, which come from long-dead biological material. When people talk about biofuels, they most often mean products such as bioethanol and biodiesel, which are derived from planted crops such as corn, sugar cane, or soybeans. But other

5. Emitted CO_2

3. Fossil fuel combusted in power stations

4. Power supplied

1. Carbon stored underground as coal seam or as crude oil or natural gas deposit

2. Fossil fuel mined, refined, processed, transported

Fossil Fuels

Fossil Fuels Versus Biofuels

Use of a fossil fuel such as coal is an "open" process. Coal—which can be thought of as a form of carbon locked securely underground—is mined and combusted to provide energy, but this involves the release of CO_2 into the atmosphere. The use of biofuels occurs in a "closed cycle." Combustion of a biofuel also pumps CO_2 into the atmosphere, but this is balanced by the previous absorption of some CO_2—

types of biomass—such as dead trees, animal waste, and biodegradable human waste—also have the potential to be used as biofuels.

In theory at least, biofuels have an environmental advantage over fossil fuels such as coal, oil, and gas (see *Fossil Fuels Versus Biofuels*, below). In practice, however, in-depth studies of the environmental costs of biofuels indicate that many are not especially environmentally friendly. For example, growing corn in the US, soybeans in Brazil, and palm oil in Malaysia as biofuels may be no more environmentally friendly than using fossil fuels. Converting land to growing these crops may also lead to damaging land-use change elsewhere.

PHOTOSYNTHESIS:
A process by which green plants and some other organisms use light energy to convert CO_2 and water into carbohydrate (a nutrient), with the release of oxygen.

Jargon buster

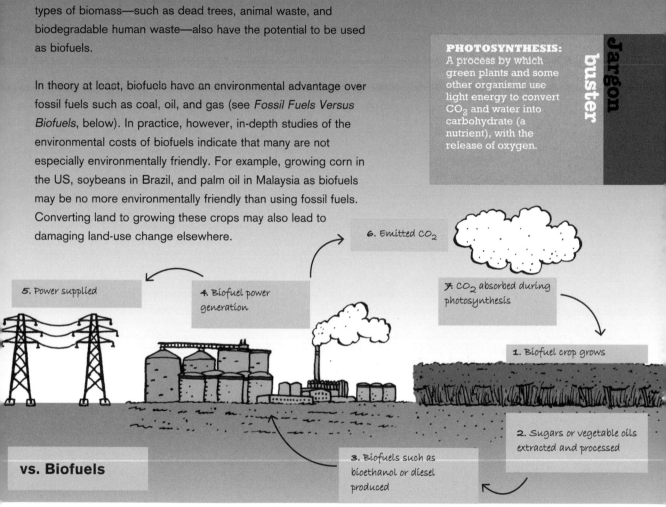

6. Emitted CO_2

5. Power supplied

4. Biofuel power generation

7. CO_2 absorbed during photosynthesis

1. Biofuel crop grows

2. Sugars or vegetable oils extracted and processed

3. Biofuels such as bioethanol or diesel produced

vs. Biofuels

through the process of photosynthesis—when the plant from which the biofuel is derived was growing.

Main Drawbacks

Some of the main issues and problems relating to the use of biofuels are:

• Converting forests, swamps, or prairies to produce biofuels can release vast amounts of CO_2 into the atmosphere, which can take decades to mitigate. Apart from the CO_2 emitted

if the wood from forest clearances is burned, additional CO_2 gas is released when biomass in disturbed soil—for example, roots—is broken down.

• Calculating exactly the volume of greenhouse gases produced in burning biofuels is complex and inexact. So projects that at first sight seem to be "carbon saving" may actually turn out in the end to be "carbon emitting."

• There is also a food versus fuel choice with biofuels. Should food-

producing farmland be diverted to the production of biofuels?

The Way Ahead

New biofuels under development, including those made from algae, grass, wood, and recycled products such as cooking oil, look likely to be more environmentally friendly than the current sources of biofuel. Moreover, planting biofuel crops on degraded and abandoned agricultural lands has a clear advantage over the destruction of forests for biofuel crops.

- Vehicles relying at least partly on electricity for power range from ordinary and plug-in hybrids to all-electric and hydrogen fuel-cell vehicles.

- Vehicles of these types are significantly more environmentally friendly than gasoline or diesel-driven vehicles, or have more potential to be.

- Electric vehicles are of most use for frequent short journeys in and around cities, but this covers a high proportion of automobile use.

- The cost of "fueling" a plug-in hybrid or all-electric car with electricity can be as little as a third the cost of fueling a gasoline-powered car.

Electric Vehicles

	Basic Features	**Pros and Cons**
Hybrids 	A hybrid vehicle contains both a battery-powered electric motor that drives the car at slower speeds and a gasoline- or diesel-fueled internal combustion engine that takes over at higher speeds and recharges the battery.	Gasoline-fueled hybrids can travel 25–30 percent farther on the same amount of fuel as gasoline-only cars. They tend to be most economical over many short trips. Since they burn less fuel, hybrids emit less greenhouse gases than regular gasoline or diesel-fueled vehicles, but the batteries are costly and there are environmental impacts from making them.
Plug-in Hybrids 	Plug-in hybrids operate in a similar way to hybrids, but with more power being provided by the electric motor. The gasoline-fueled engine is utilized only when the battery charge is low. At present, their electric-only range is usually no more than about 12 miles.	A plug-in uses much less gasoline or diesel than a comparable ordinary hybrid. Against that, there is the cost of having to regularly recharge the battery, but this is generally far less than the amount saved from reduced purchases of regular fuel. Plug-in hybrids require bulky, costly batteries; the frequent need to recharge can be inconvenient.
Electric Vehicles 	All-electric vehicles run solely on a rechargeable battery that can power the vehicles for about 140 miles, although one all-electric sports car now in production can travel for more than 240 miles on its lithium-ion battery pack.	All-electric vehicles are cheap to fuel. The vehicles are non-polluting, although production of the electricity used to recharge them may have an environmental impact. The battery—usually so large the car has to be designed around it—is expensive to replace. Many regions or countries have, as yet, few recharging stations.
Hydrogen Fuel-Cell Vehicles 	These vehicles are driven by an electric motor powered by a stack of hydrogen fuel cells (a device similar to a battery). It works by converting hydrogen (required as a fuel) and oxygen (taken from the air) into water, producing electrical energy as a by-product.	Hydrogen fuel-cell vehicles emit no pollutants. By weight, hydrogen contains more stored energy than gasoline, which implies greater distances traveled between refills. If the technology becomes common, the cost of hydrogen fuel is likely to drop below that of gasoline. However, the cheapest source of hydrogen is natural gas, a fossil fuel, and hydrogen extraction emits high levels of greenhouse gases.

Rising gasoline prices, together with the urgent need to reduce greenhouse gas emissions, have in recent years rekindled interest in cars and other vehicles that are powered either partly or completely by electricity instead of gasoline. A range of different types of vehicle with an electrical power component is currently available, from regular hybrid cars (which are still basically fueled by gasoline) to plug-in hybrids, all-electric vehicles, and hydrogen fuel-cell vehicles.

LITHIUM-ION BATTERY: A type of rechargeable battery with a high energy-to-weight ratio and slow loss of charge when not in use. These batteries rely on lithium ions that move in one direction through the battery during discharge and in the opposite direction when being charged.

Jargon buster

Outlook and Main Technical Challenges

Hybrid vehicles have proved popular among environmentally conscious consumers, and are encouraged in many countries by tax incentives. They are likely to have an important place in the vehicle market for the foreseeable future. Many different models are now available in most developed countries, including sedans and pick-up trucks.

Several major car manufacturers are planning to introduce plug-ins by around 2012. The main technical challenge concerns the battery. Nickel-metal hydride batteries used in ordinary hybrids don't hold enough energy to make them viable for plug-ins. Lithium-ion batteries hold more energy, but cost more to make and have high operating temperatures.

So far, few affordable, robust all-electric cars have been produced, but a number of major manufacturers are expected to introduce models over the next few years. The main challenge is to make a battery that is smaller, more powerful, safer, and less expensive. They operate quietly on cheap fuel, with zero emissions—a popular choice for the future.

The main challenge to the development of these vehicles is the building of an infrastructure to extract and store compressed hydrogen in large quantities and to deliver it to road users. This is likely to cost billions of dollars—at present there are virtually no such facilities, and hardly any hydrogen filling stations.

bite size facts

• It is estimated that recycling efforts in the US alone reduce net carbon dioxide emissions by about 50 million tons a year.

• One recycled aluminum can saves the energy it takes to run an average-sized TV for three hours.

• Paper loses quality each time it is recycled, whereas glass can be recycled indefinitely without any deterioration in quality.

• The world's most recycled materials are iron and steel; any grade of steel can be recycled to high quality new metal.

Recycling

What is the point of recycling? As many people know, one of the main reasons is to save on energy—and since energy most often comes from burning fossil fuels, this helps reduce carbon emissions. These energy savings can be quite significant. For example, one recycled glass bottle can save enough energy to power a computer for 25 minutes, while one recycled plastic bottle can save enough energy to power a 60-watt lightbulb for three hours. Recycling paper uses about 70 percent less energy than making it from raw materials, while recycling aluminum produces a 90 percent saving. There are many positives to recycling: it generates less physical waste, and so less of the landscape has to be devoted to landfill sites; it reduces environmental pollution, a common result of landfills; and it can slow down the consumption of scarce physical resources such as minerals and forests.

GOVERNMENT SUBSIDY

RAW MATERIALS PRICES

RECYCLED MATERIALS PRICES

LANDFILL COSTS

"Recyclonomics"

The economics of recycling are complex and vary from one product to another and from one region of the world to another. They include factors such as the relative market prices of recycled and nonrecycled materials, the costs of landfill sites, the energy and financial costs of separating different types of waste, the extent to which governments will subsidize or legislate to encourage recycling to reduce greenhouse gas emissions, and many other factors. These factors affect what large-scale recycling opportunities are available in different regions—in other words, what can and cannot be recycled. In most areas of most developed countries, paper and cardboard, glass, metals, car batteries, aluminum cans, plastic containers, and steel from old buildings are now all commonly recycled in large quantities. Other materials that can be recycled include concrete aggregate, paint, textiles and timber, and green waste (grass cuttings, hedge trimmings) for conversion into fertile topsoil by composting.

To Recycle or Not to Recycle?

Although obviously it is not environmentally friendly to drive several miles to a recycling center only to deposit a small number of items, where the opportunity to recycle a material is available, there is usually a clear environmental benefit to participating in a recycling operation. The amount of the environmental benefit does, however, vary with the material in question. For example, recycling aluminum cans is particularly environmentally friendly in terms of the savings in both energy and air pollution. Recycling of glass, paper, and other metals have less dramatic but still significant energy savings, while the advantages of recycling some plastics are more marginal. Where objects or materials cannot be recycled—for example, lightbulbs, ceramics, plastic bags, styrofoam packaging—this is usually because the energy costs of separation and/or reprocessing outweigh any other benefit, environmental or otherwise.

RECYCLE?

YES   NO

COSTS OF WASTE SEPARATION

LABOR COSTS OF RECYCLING

ENERGY COSTS OF RECYCLING

7 Health Stuff

CHAPTER CONTENTS

This chapter looks at the science surrounding some common issues in human health, many of them to do with nutrition, drugs and medicines, and the risks to health of such diverse phenomena as flu viruses, cellphone usage, and excessive sunbathing.

Science and Health

Science as applied to human health is often referred to as medical science. Some of its major branches include biochemistry (the study of chemical processes in the body), pharmacology (the study of drugs), and human genetics (how genes affect health), but there are also many others such as epidemiology, which looks at the different factors affecting the health of populations. The fruits of research over the past 80 years include the development of many new and effective vaccines and antibiotics to combat infectious diseases, drug treatments for many common disorders such as asthma and diabetes, and many new surgical techniques. Many challenges remain, however, including the appearance of antibiotic resistance in- some "superbugs," and the epidemic of obesity that has recently gripped much of the developed world.

Bugs, Drugs, and Behavior

There are many influences on health, ranging from factors that are entirely outside our control, such as our genes, to those we have complete control over, like what we eat, the amount of exercise we take, and the extent to which we indulge in, or avoid, harmful behaviors. First up, this chapter considers diet-related issues such as overindulgence in alcohol, the hazards of raised blood cholesterol, and obesity. Next we look at two of the main categories of "bugs" that cause illness in humans—bacteria and viruses—the latter with a particular focus on influenza pandemics. One of the most effective methods that science has devised to combat such bugs, notably vaccination, is considered next, while on the following pages, we examine some of the factors to be taken into account when assessing health stories in the media, using the suggested health risks from cellphone usage as an example. The final sections consider the possible health consequences of sunbathing and how they can be avoided, drug abuse in sport, and the controversial issue of animal testing of new medicines.

bite size facts

• The main influences on health are genetic make-up and the choices that we make in our behavior, such as what we eat and how much we exercise.

• Of five major health risks identified in developed countries, two of them—obesity and a high level of cholesterol in the blood—are diet-related.

• Two of the other major health risks—tobacco smoking and drinking too much alcohol—both have a strong addictive aspect.

• The fifth major risk to health is high blood pressure, which has a mix of causes, but is treatable.

Influences on Health

Card games are a mixture of luck and judgment. You can't choose the hand of cards from the dealer. Picking the best face-down cards is also mostly luck, although after a while you might figure out some of them. However, you can choose face-up cards and how you play the game. In a similar way, good health is also a mix of chance and choice. You can't alter what you are dealt—your genes, inherited from your parents. Faulty genes cause conditions such as cystic fibrosis or hemophilia, or increase the likelihood of problems such as heart disease, some types of cancer, diabetes, and mental conditions such as depression.

Face-Up Health Cards

Face-up health cards are the ones we can easily act on. Good cards include regular exercise, eating healthy foods, controlling your weight, and medical measures such as getting check-ups, screening tests, and vaccines. Bad cards include using harmful or addictive substances such as tobacco and too much alcohol, or eating too much food containing animal fats. Other bad cards include risky behavior such as excessive sunbathing, unprotected sex, and indulging in extreme sports without adequate prior training and precautions. And in all of this, mind your mind: keep a good work–life balance, get enough sleep, and exercise your brain as well as your body.

Our Genetic Make-Up

The cards we are dealt in our hand—our genetic make-up—cannot be altered. However, knowing about a possible genetic problem can help to reduce its effects.

Face-Down Health Cards

Your face-down health cards can be good or bad. Although the bad ones are often outside your control, if you have sufficient will and resources, you can reduce their effects. These cards include unhealthy living conditions, exposure to germs (harmful microbes), environmental pollution (such as poor air quality), and being in the vicinity of tobacco smoke.

Make the Best of It

Whatever health cards we receive, we can improve them through our own actions, choices, and behavior. The prize is a winning hand of a healthy, happy life.

The scientific evidence that smoking is bad for health is overwhelming. It causes cancers, heart disease, and many other serious conditions. Most of us know someone who smoked and lived to a good old age. But that's notable because it's so rare. On average, each cigarette shortens life by about 10 minutes—around twice the time taken to smoke it.

- The resting body uses energy at the same rate as a standard 100-watt lightbulb.

- Thinking uses a lot of energy—the brain consumes one-fifth of all the energy used by a body at rest.

- The recommended daily limit of salt is 6 grams (⅕ oz) yet some ready-made meals contain three times this amount.

- If a food label lists "sodium" instead of "salt," multiply the amount of sodium by 2.5 to get the quantity of salt.

Nutrition, Diet, and Obesity

Food is fuel. Like any machine, the human body needs fuel for energy—to move around, keep warm, and to power the heartbeat, breathing, digestion, and even thinking. But food is more than fuel. It also supplies nutrients and raw materials for growth, maintaining and repairing body tissues, and making sure all internal processes work smoothly. So what you eat is as important as how much.

This leads to the idea of the food pyramid. It shows foods we should consume in large amounts, those to limit, and those to avoid. A "balanced diet" means eating the correct amounts, in proportion, of a range of dietary components.

Eat Like This...

Follow the healthy eating pyramid on this page and your body will always be thankful. The unhealthy eating pyramid on the opposite page would be a disaster.

Put a limit on foods high in fats, oils, and sugars, such as chocolate. They are "empty calories" with too much energy and few other nutrients.

Eat some high-protein foods such as dairy products, legumes, and (for non-vegetarians) meat and fish, but not too much. These foods contain essential nutrients for body maintenance.

The bulk of your intake should be cereal products such as wholegrain bread, vegetables, and fruits. These provide energy, vitamins and minerals, and dietary fiber.

AMINO ACIDS: Small chemical units that join like links in a chain to make proteins.

PROTEINS: Varied substances that form the main structural parts of body tissues such as muscles, bones, skin, and hair.

Carbohydrates provide energy. They are found in starchy foods like grains, breads, pasta, and potatoes. Proteins supply amino acids for tissue growth, maintenance, and repair. They are mainly found in meat, fish, dairy produce, and legumes. Fats and oils are needed for body tissues like nerves, but only in small amounts. Plant oils are healthiest. Vitamins and minerals are required for hundreds of internal processes; among the best sources are cereals and cereal products, and fresh fruits and vegetables, all of which also supply dietary fiber. Foods of animal origin also supply many vitamins and minerals. Last and far from least is water—in general at least two to three quarts (or liters) should be drunk each day.

Don't Eat Like This...

Too many sugary foods such as cookies and cakes give a "rush" of energy that soon fades. The body prefers the slow-release, complex carbohydrates found in wholegrains, potatoes, and other starchy vegetables.

Alcoholic drinks provide calories (energy) from their alcohol and carbohydrate content, but little or no other nutrients. High consumption of alcoholic drinks is a frequent contributor to weight gain.

Animal fats and oils in red meats, processed meats such as salamis and burgers, fried foods, and cheeses are linked to high cholesterol, obesity, and heart disease, as explained on the next page.

The Diabetes Iceberg

The hormone insulin, made in the pancreas gland behind the stomach, controls how microscopic cells take in their main energy source—glucose, a form of sugar. In diabetes, the body cannot use glucose in the usual way. In type 1 diabetes, insulin production fails, while in type 2, cells do not respond normally to insulin.

In both forms of diabetes, glucose builds up in the blood and causes problems around the body, from the kidneys to the muscles and eyes, and increases susceptibility to infection. Type 2 diabetes is closely linked to obesity. Cases of both types are increasing fast in richer countries, and even faster among children. Yet they are only the tip of an iceberg. Many people have type 2 diabetes that has not yet been diagnosed by a doctor. Even more common is "pre-diabetes," in which a person is heading toward type 2 without realizing it.

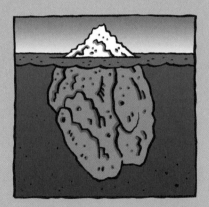

Above the surface of the diabetes iceberg are people who know they have the condition. Below the surface are many more who have diabetes but don't know it, or who have prediabetic conditions. It's an enormous health time-bomb.

The Dangers of High Cholestorol

Cholesterol is a "building block" body chemical for the outer skin-like membranes of the body's trillions of microscopic cells. It is also needed to make the digestive fluid bile and certain hormones, such as the female sex hormone estrogen and epinephrine, a hormone involved in the "fight-or-flight" response to danger. Too much cholesterol, measured in the blood, can cause various problems—primarily fatty deposits, known as atherosclerosis, in the linings of the arteries. Narrowed arteries can become blocked and increase the likelihood of the formation of blood clots. Blood clots can lodge in other vessels, possibly causing a stroke or heart attack. Increased blood cholesterol is often due to eating too much fatty food, especially saturated fats from animal sources. The medical advice is clear: Cut back on fatty meats and other animal-based fats and oils.

Alcohol: How Much is Too Much?

In the US, alcohol intake is measured in standard drinks (each containing about 0.6 fluid oz. of alcohol). Recommended weekly limits are 14 standard drinks for men and 7 for women. This difference is partly due to the size difference, but also because male and female bodies handle alcohol slightly differently. Regular drinking beyond these limits can lead to many serious health problems affecting the brain such as memory loss and even dementia— also higher risk of heart disease, stroke, liver problems such as cirrhosis, breast cancer, ulcers, and other digestive system damage.

How Many Units?

1   a regular bottle or can (350 ml) of medium strength beer, a small glass (110 ml) of wine, or a single shot of liquor

1.5   a regular bottle or can of full strength beer, pre-mixed spirits or an alcopop; or a large glass (165 ml) of wine

7   a typical bottle of wine (750 ml)

22   a 700 ml bottle of liquor

ARTERY: A vessel or tube that carries blood away from the heart.

HORMONES: The body's internal chemical messengers and coordinators, controlling many processes.

EPIDEMIC: Many more cases of an illness or disease than is normal in a given period of time, usually occurring over a widening area.

Jargon buster

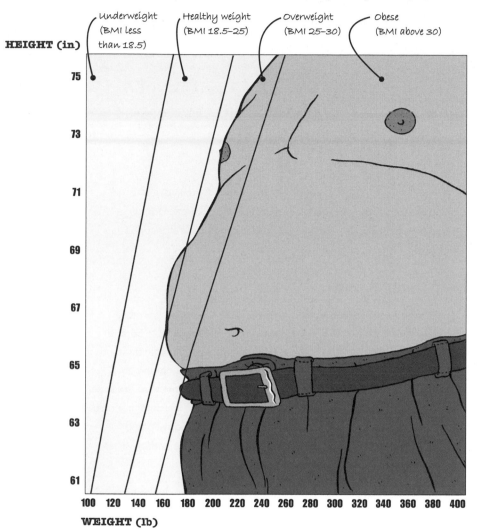

HEIGHT (in)

Underweight
(BMI less
than 18.5)

Healthy weight
(BMI 18.5–25)

Overweight
(BMI 25–30)

Obese
(BMI above 30)

75

73

71

69

67

65

63

61

100 120 140 160 180 200 220 240 260 280 300 320 340 360 380 400

WEIGHT (lb)

Recognizing the Signs of Obesity

If you eat too much, the body converts the excess food to fat. Body Mass Index
(BMI) is a measure of obesity. It's calculated by multiplying the body weight in
pounds by 700 and dividing the result by the height in inches multiplied by
itself: lb x 700 ÷ (in x in)*. Someone who weighs 150 lb and is 70 inches (5 ft,
10 in) tall has a BMI of (150 x 700) ÷ (70 x 70), which is 21.4—very normal.
A BMI of more than 25 usually means a person is overweight, and one above
30 is a sign of obesity. This brings an increased risk of high blood pressure,
heart disease, diabetes, arthritis, and many other health problems. The remedy
is simple: eat less and exercise more.

* In metric units, divide the body weight in kilograms by the height in meters multiplied by
itself: kg ÷ (m x m)

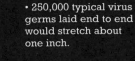

- 250,000 typical virus germs laid end to end would stretch about one inch.

- 25,000 average-sized bacteria laid end to end would stretch about one inch.

- 5,000 typical protist (protozoan) germs laid end to end would stretch about one inch.

- 2,500 typical fungus spore germs laid end to end would stretch about one inch.

Bugs and Superbugs

There are many kinds of "bugs"—from insects such as fleas and lice to various microscopic life-forms. There are four main kinds of micro-bugs or germs that cause many kinds of infectious diseases. These are described in the panel below.

Drugs to Combat Bugs

Our main weapons against bacteria are antibiotic drugs, for which a more accurate name is antibacterials. The first one, discovered back in 1928 and still widely used, is penicillin. Some antibiotics interfere with the way bacteria split in two as they multiply. Others affect the outer skin or membrane and cause bacteria to clump together. Antibiotics are useless against viral illnesses, so there is no point taking an antibiotic to treat, for instance, the common cold, which is a viral infection. However, some serious and rapidly developing illnesses, such as meningitis, can be caused by either bacteria and/or viruses. Until a firm diagnosis is made, a range of drugs may be tried.

There are also drugs that act against viruses, but compared to antibiotics, the range is much more limited and sometimes less effective. Our main weapons against viruses are vaccines, as explained on pages 134–135.

Flavors of Penicillin

Today, bacteria are more often fought with a chemical derivative of penicillin rather than the original penicillin. Derivatives are created by making small changes to the penicillin molecule and have names that still end in "-cillin." They include such antibiotics as amoxicillin, methicillin, and flucloxacillin.

Microbugs: The Big Four

- LARGEST are fungi or molds. They cause athlete's foot (tinea), thrush (candida), aspergillosis in the lungs, and fungal nail infections.

- NEXT LARGEST are wriggly single-celled protists (or protozoa), such as plasmodium, which cause malaria, and trypanosomes, which produce sleeping sickness.

- EVEN SMALLER are bacteria, also single cells, which are about 100 times smaller than the body's own cells. They cause a huge range of infections such as anthrax, tuberculosis, cholera, gonorrhea, syphilis, diphtheria, whooping cough, typhoid, skin infections such as boils and impetigo, stomach ulcers, and many cases of food poisoning and diarrhea.

- TINIEST OF ALL, a hundred times smaller than bacteria, are viruses. They are responsible for various kinds of colds and flu (influenza), cold sores and herpes, chickenpox, measles, rubella (German measles), rabies, AIDS, hepatitis, yellow fever, cervical cancer, and many more.

Superbugs and How They Multiply

As living things multiply, slight changes or mutations may occur in their genetic material. Bugs number trillions and some can multiply from one to 1,000 in just a few hours. These vast numbers mean mutations are very likely to occur. Sometimes a purely chance mutation gives a bug partial or complete resistance to an antibiotic or antiviral drug. This new, drug-resistant strain is now a "superbug." It can multiply fast—until medical scientists find a new drug to kill it. One example is MRSA. It's a strain of the common bacterium Staphylcoccus aureus (SA) that is resistant (R) to the group of antibiotics based on methicillin (M). Taking antibiotics that are not really needed, such as for a viral illness, or not finishing a course of antibiotics properly, is a waste of their power and can encourage the emergence of these "superbugs."

Jargon buster

INFECTIOUS: Caused by some kind of microbe or similar living thing and likely to be spread.

CONTAGIOUS: Able to be passed from one sufferer to another by direct contact, shared objects, or body fluids.

DNA: Deoxyribonucleic acid, a substance that contains instructions called genes for a living organism.

From One to a Million in Seven Hours

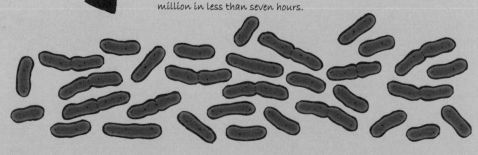

1. Most bacteria can multiply simply by splitting themselves in half, a process called binary fission.

2. After two divisions there are four "daughter" bacteria, after three divisions there are eight, and so on.

3. The genetic material is copied every time so each "daughter" receives a complete set. However occasional mistakes produce mutant bacteria that can live in different conditions.

4. In favorable surroundings some bacteria undergo fission every 20 minutes, resulting in more than one million in less than seven hours.

• Every year "regular" or "seasonal" flu kills up to 40,000 people in the US and as many as 4,000 in the UK.

• One of history's biggest pandemics was Spanish flu in 1918–1920. It killed up to 80 million people worldwide.

• The H1N1 swine flu virus is a new strain of virus containing parts of several different viruses that had previously affected pigs, birds, and humans.

• Influenza viruses are most often spread by coughing and sneezing or by touching contaminated surfaces and then the nose or mouth.

Flu Pandemics

A pandemic is bigger than an epidemic. It is an outbreak of an infectious disease that affects thousands, perhaps millions, over an area as great as a continent or even much of the world. Historical pandemics included bubonic plague, tuberculosis, smallpox, and more recently HIV/AIDS. There are also regular flu (influenza) pandemics caused by viruses called orthomyxoviruses. Some of these spread from animals to people. Viruses can do this because as they multiply, changes occur in their genetic material, known as mutations (see pages 130–131). A virus strain well-adapted to infecting a certain animal may, with a new mutation, "jump species" and be able to attack other cells such as human body cells.

How a Flu Virus Multiplies

1. A typical flu virus looks like a golf ball with tiny knob-ended stalks all over it. Within this casing is its genetic material, DNA. It is always ready to multiply.

6. The cell bursts open to release many viruses, which go on to infect other cells. The whole process may take less than 15 minutes.

Bird Flu (Avian Influenza)

This type of flu first "jumped" to humans around 1997. It has four main virus strains, the most worrying one being H5N1. It affects only those humans in intimate contact with infected birds, usually poultry, as can occur if the bird's fluids or droppings enter the human body through the mouth, nose, or a cut. Unlike swine flu (see opposite), cases of human-to-human transmission are not yet known, so bird flu is still relatively rare and isolated. Even so, it's very serious. It produces the usual flu-like symptoms (fever, shivering, headache and joint aches, sickness, and sore eyes, nose and throat) and kills up to two-thirds of sufferers. Symptoms can be eased with antiviral drugs. If the virus does one day mutate into a form that spreads between humans, it could start a pandemic.

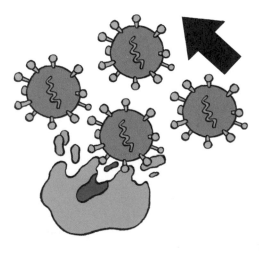

Swine Flu

The first cases of the flu pandemic known as "swine flu" were diagnosed in Mexico and the US in early 2009. The virus—flu type H1N1—may have circulated in pigs for years before crossing to people (it's not caught from pig products). Unlike bird flu, swine flu can spread directly from person to person, making it much more common, with thousands of new cases reported weekly in dozens of countries in the fall of 2009. However, it kills fewer than one in 250 sufferers, mainly people with underlying health problems. Vaccines against it were developed quickly.

Bug Names and Strains

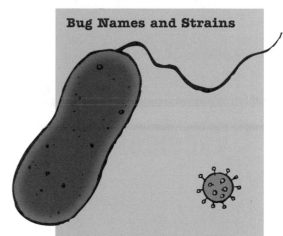

Bacteria are hundreds of times bigger than viruses. Some have long "tails", *flagella*, which they twirl to move along. They are also truly living, with inner processes similar to those of other cells, including our body cells. Not all bacteria are dangerous—trillions live harmlessly in the soil, where they help to decompose dead plant and animal matter. Harmful types of bacteria are known as pathogenic. There are hundreds of kinds or species, each with many strains, such as the "superbug" MRSA (see page 131).

All viruses are pathogenic (disease-causing) since they must take over a living cell to multiply, destroying this host cell in the process. Viruses also exist in many strains, often with strange-sounding code numbers and letters. These are based on how they were identified or isolated, or their structural parts. In the feared H5N1 strain of influenza A virus, H5 means it is the fifth version discovered of chemical subunits (proteins) in its outer casing called hemagglutinins. These attach the virus to its host cell. N1 means it has the first version identified of another protein, neuraminidase, which helps new viruses to leave their dying host.

2. To reproduce, a virus needs a living cell, such as a bacterium or a body cell in an animal or plant. It attaches to the cell's outer skin or membrane, and makes a hole in it.

3. The genetic material of the virus passes from the virus into the living cell. Here it begins to take over the cell's internal processes.

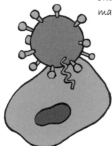

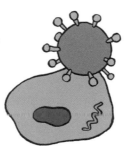

5. The cell's own life processes are disrupted, it runs out of energy and nutrients, and begins to die. Each length of viral genetic material forms its own outer casing.

4. The cell's "production machinery" makes hundreds or even thousands of identical copies of the viral genetic material.

bite size facts

• Vaccination was first shown to be successful by English country doctor Edward Jenner in 1796.

• In 1979 the World Health Organization declared the terrible disease of smallpox completely wiped out, thanks to vaccination.

• A vaccine against malaria would save up to three million lives and prevent up to 500 million from suffering the disease each year. Researchers have been looking for a vaccine against the disease for almost 200 years, but a safe, effective one is still some years away.

Vaccines

A typical vaccine contains killed or weakened forms of a harmful microbe, or germ. It is put into the body by means of an injection, or in some cases by tablet or spray. Soon the body's immune system, which provides its natural defenses, swings into action. The system recognizes the germs as foreign and harmful, and destroys them using various kinds of white blood cells and substances known as antibodies. However, in contrast to what happens when real germs invade, the body suffers no illness. However, after the initial battle, the immune system develops "memory" cells for that particular germ. In the future, if exposed to real disease-causing germs, these memory cells identify them so

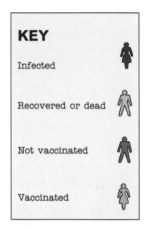

KEY

Infected

Recovered or dead

Not vaccinated

Vaccinated

Benefits of Mass Vaccination

A vaccine does not just prevent a disease in the people who receive it. The vaccine also protects others. This is known as "herd immunity." Typically, if 90 percent or more of the people in an area are vaccinated and become immune, epidemics can never take hold. In the two illustrated sequences (right), the progress of a disease in a community with no vaccination (top) is compared with that in one where many are vaccinated (below). Red people have an infectious disease, dark blue people are not vaccinated or immune, light blue people are vaccinated and immune, and yellow people have either died or recovered and have developed natural immunity.

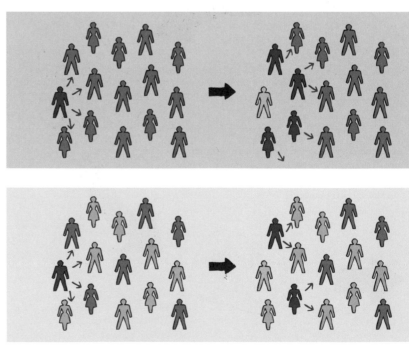

quickly that the immune system kills them before they multiply sufficiently to cause any illness. This process is known as immunization. Giving the vaccine, called vaccination, makes people immune or resistant to that disease.

Some vaccines contain only parts of a germ, such as a section of its outer covering, which is enough to stimulate the immune system. Other vaccines work against harmful chemicals, called toxins, produced by the germ.

Vaccines are available against many viral and some bacterial diseases. Their great successes include conquering smallpox, and vastly reducing polio, measles, diphtheria, tetanus, and some types of bacterial meningitis. Most vaccines are given to babies and young children, so that their immune systems can gain strength before they encounter the real germs by chance. Some others are recommended for travel to particular countries. For convenience, several vaccines are often given together in a combined form.

Vaccine Scares

Highlighting a vaccine's extremely rare but serious adverse effects can not only cause mass panic, but also unnecessary suffering and death. For example, measles vaccination produces a serious brain inflammation in one case in a million. But this is not a reason to avoid vaccination, since the risk of dying from measles is some 3,000 times higher. A recent UK scare suggested a link (now disproved) between MMR (mumps, measles, rubella) vaccine and autism. Uptake of the vaccine fell, allowing more cases of mumps and measles, with some severe disabilities and a few deaths occurring as a result.

Experts continually strive to improve vaccines in order to reduce adverse effects. Medical records are also constantly analyzed for links between health conditions and vaccine complications. This identifies at-risk people, to whom special advice is then given rather than the vaccine.

UNVACCINATED COMMUNITY
At each stage of the disease, the germs multiply in non-immune people, who then spread them onward by touch, or in tiny droplets they breathe out, sneeze away, or cough up. The germs keep breeding to infect others.

VACCINATED COMMUNITY
With plenty of vaccinated people, the germs have far fewer bodies in which to multiply. Also, people who catch the disease naturally become immune. Soon the germs have nowhere left to breed and the infection fizzles out.

- If you value your health and travel a lot, take public transport. By far the safest, in order are air, rail, bus, and ship.

- The safest forms of private transportation are car, bicycle, then by foot (in that order), and riskiest by far is motorcycle.

- Age is an inescapable risk to health. More than three-quarters of strokes occur in people over 65 years of age.

- The world's oldest-ever person, Jeanne Calment of France, who lived to 122 years and 164 days, loved olive oil, chocolate, and port wine.

Relative Health Risks

Almost every day we hear news about what is good and bad for our health. A particular food or drink, a medicine or pill, going on a diet, or playing this sport or doing that exercise... the list could fill this book. But what is really harmful to your health? How can we put risks into some sort of perspective? Earlier in this chapter the "Big Five" risks to health were listed—smoking, obesity, high blood cholesterol, high blood pressure, and drinking too much alcohol. For the average person in reasonable health, these far outweigh all other risks combined. Look at these risks from another perspective and you have five major ways to stay healthy—don't smoke, eat sensibly to maintain a healthy weight, control blood cholesterol, monitor your blood pressure, and don't drink to excess.

Comparing the Risk

The graphic on the right shows some relative risks for people living in an average developed country. Note how relatively uncommon are the tumors that have been linked (in some studies) to long-term cellphone use. Even if the incidence of these tumors were to double, the number of new cases would still be less than the number of nonsmokers getting lung cancer in one year and far less than the number of smokers getting lung cancer. Note also how uncommon deaths from plane crashes are compared to deaths from road accidents.

Take a million people in an average developed country over the course of a year:

About 2,000 die of heart disease

About 1,000 are involved in a serious road accident

About 600 get lung cancer and are smokers

About 100 die in a road accident

About 80 get lung cancer and are nonsmokers

About 40 develop an acoustic neuroma or a glioma (two types of brain tumor)

About 1 dies in an aviation accident

Of course, health risks are affected by many factors. Age is the main one; also being male or female, where you live, your job, leisure activities, and in some cases your ethnic group. For males aged 15–30 years in poorer parts of big cities, the main health threats stem from driving cars or motorcycles, taking drugs, and from crime. As you get older, risks steadily increase from the "Big Three" of cancers, heart disease, and stroke.

Are Cellphones Dangerous?

An example of how headline stories about health risks need to be considered in perspective concerns cellphones. When held to the ear, a cellphone's radio signals cause a slight heating of nearby tissues in the brain. Early surveys showed no hazard to health. One or two recent studies have suggested that long-term mobile phone use (more than 10 years) may carry some increased risk of tumors, including acoustic neuromas (curable) and gliomas (noncurable). But these studies are by no means accepted as conclusive—other studies have found no increased risk. Furthermore, the hazard, if it exists, is one that raises an extremely small existing risk of a tumor to one that is merely very small.

Tumor Causes
The exact causes of most of these uncommon to rare tumors of the nervous system are not known. A few are thought to be the result of inherited (genetic) disorders. Acoustic neuromas are not life-threatening and can often be successfully treated.

A much greater problem is that of using a cellphone while driving, whether hand-held or hands-free. It increases by about four times the chances of an accident, with injury or death. This risk is starting from a much higher level than that of the risk of a brain tumor, since road traffic accidents are already a 1-in-1,000 hazard each year for most age groups. Consequently, the use of a hand-held cellphone while driving is now banned in many countries.

Surveys and Sample Size
When a survey is based on a very small sample size, chance plays a disproportionately large part. Imagine flipping a coin five times with a microwave switched off, and then with it switched on. Chance could give four heads when off and only two when on. With many more flips these numbers would eventually average out. But from the tiny sample in the first survey, it seems that switching a microwave on means a coin is less likely to come up heads. If you were to rely on these results, you could start looking for a cause, even though it is the "study" that is essentially unreliable.

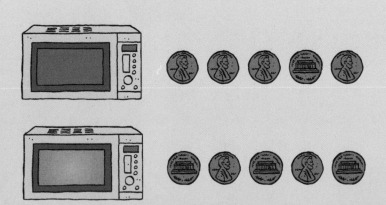

• Excessive exposure to sunshine can have various harmful effects on the skin, which derive mainly from the ultraviolet (UV) components of solar radiation.

• Long-term exposure to soot, pitch, asphalt, coal tar, creosote, or paraffin wax can increase the risk of the non-melanoma forms of skin cancer.

• Malignant melanoma accounts for only one case of skin cancer in 25, but it causes four out of five deaths due to skin cancer.

• When selecting a sunscreen, it is important to choose a product that protects against both UV-A and UV-B.

Sun and Skin

Some people see suntanned skin as attractive, sexy, healthy, and perhaps a sign of having vacationed somewhere hot and sunny (which could mean the person wearing the tan has money and leisure time). Others view a suntan as wrinkled and saggy skin and/or skin cancers waiting to happen. The complicated reality is somewhere inbetween. There are many factors involved—not least, the Sun Protection Factor (SPF) labeling on the sunscreen you buy. The rays in sunlight that affect skin most are called ultraviolet, UV, with two main kinds, UV-A and UV-B.

Ultraviolet-A

UV-A is the main cause of a suntan, encouraging the skin to manufacture more of its natural coloring substance, melanin. After this happens, a heavily suntanned person, along with those who naturally have very dark skin, are better protected than light-skinned people against UV dangers. UV-A is linked to premature skin aging (as is UV-B) and also the development of malignant melanoma, a rare but sometimes fatal form of skin cancer.

Ultraviolet-B

UV-B is the main cause of sunburn (as distinct from suntan) and produces wrinkles and other indicators of premature skin aging. It is also linked to the development of common but treatable skin cancers such as squamous cell carcinoma and basal cell carcinoma. Because of this, some sunbed and tanning salon lamps produce more UV-A than UV-B. But both forms of UV radiation are now linked to skin cancers, so these lamps are best avoided.

When is UV at its Strongest?

The main factor affecting UV intensity is the Sun's height in the sky, so UV levels are higher the closer you are to the equator. They are also higher in summer than in winter, and from 11 A.M.–3 P.M. rather than earlier or later in the day. UV reflects off snow, and its intensity increases with altitude, affecting skiers and mountaineers. UV also partly reflects off water onto those doing watersports above the surface of the water.

Why Some Sun is Healthy

Small amounts of UV-B are beneficial for the body. They help the skin to produce vitamin D, which is important for healthy bones, the body's self-defense immune system, and for anti-inflammatory processes in the body. Vitamin D also has a protective effect against several forms of non-skin cancers, including breast, ovarian, pancreatic, and colon cancers.

Do Suncreens Provide Protection?

Sunscreen products usually contain one or both of two main types of active ingredients. The first type is a physical sunscreen, such as zinc oxide, which works by reflecting both UV-A and UV-B away from the skin. The other is a chemical sunscreen that absorbs and dissipates the energy of UV-B, or less commonly UV-A.

Most sunscreens are labeled according to the Sun Protection Factor (SPF) they offer. This indicates how effectively they block UV-B. But it does not show overall protection, since it does not take into account how well the product deals with UV-A. Although sunscreens protect against sunburn, and may help to reduce some types of skin damage, there is no clear evidence that they protect against malignant melanoma. In fact some scientific studies show a link between sunscreen use and a higher risk of melanoma. A possible reason is that many sunscreens provide little protection against UV-A, so users may have a false sense of security, thinking that they can stay longer in the sun without realizing that UV-A is causing serious damage. For this reason, when choosing a sunscreen it is important to check that it protects against both UV-A and UV-B.

- In the 1904 Olympics, the winner of the marathon did so with the help of a small dose of a stimulant given during the race.

- Over 200 different drugs are now banned during the Olympics, including more than 50 stimulants and more than 40 different anabolic steroids.

- Some side-effects of taking anabolic steroids are, in men, testicular shrinking and breast enlargement; and in women, a deeper voice and more facial hair.

- Cycling authorities now routinely suspend from racing any athlete found to have an abnormally high concentration of red blood cells.

Drugs and Sport

The use of performance-enhancing drugs in sports cheats fellow competitors, endangers health, and risks embarrassment, a ban from competing, disgrace, and condemnation. Why do people do it? To improve their own achievements, beat a particular rival, gain fame by winning medals and championships, and get rich from awards, sponsorship, and advertising. Drug or doping cheats hit the headlines regularly, especially in athletics and cycling, but also in many other sports—even archery.

A Steadying Hand

The drugs that have most often been abused in sports such as archery and shooting belong to a group called beta blockers. These drugs act mainly on the pumping action of the heart, but one of their side-effects is to reduce muscle tremor—of obvious benefit in events where precise control over muscle movements is needed.

Catching the cheats is a constant battle. Drug-testing methods include random blood and urine analysis during or near the time of competition, and random searches for the drugs or their equipment. Authorities must continually develop new tests to detect ever-sneakier doping methods. Rarely a drug-taker owns up and becomes "clean"—that is, no longer uses drugs. This can also expose the latest tricks of the doping business, giving the authorities a boost to their battle.

Performance-Enhancing Drugs

Several types or classes of performance-improving drugs are used in sports. A linked problem is so-called recreational drugs that may not increase performance, but which may be banned near or during competition by the sport's rules or even by law.

STIMULANTS
Effects: Increased alertness, reaction time, and body processes; delayed fatigue.
Main risks: Lack of concentration, mood swings, and sleeplessness.

ANABOLIC STEROIDS
Effects: Increase the bulk of various tissues, used especially to build muscle.
Main risks: Heart, liver, mental, and skin problems, sterility, masculinization in females.

MASKING DRUGS
Effects: Taken to prevent other drugs being detected in the body.
Main risks: Fluid imbalance, dehydration, infections.

Blood Doping

"Doping" refers to drug-taking in general. "Blood doping" describes methods of increasing the level of red cells in the blood. The presence of extra red cells means the blood can carry more oxygen to the body's muscles, including the heart muscle, allowing them to work harder and longer. One method is to take out a quantity of blood, allow the body gradually to replace the red cells, then to replace the previously removed red cells. Another is to inject doses of the natural hormone erythropoietin (EPO), which stimulates bone marrow to manufacture red cells. However, a high concentration of red cells in the blood increases the risk of strokes and heart problems.

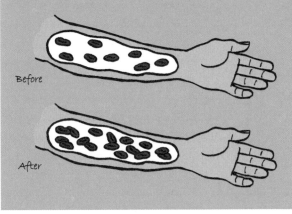

Before

After

More Muscle

Anabolic ("building") steroid drugs are substances related to the natural male hormone testosterone. Most but not all are synthetic substances. Popularly known as "juice," examples include androstenedione ("andro") and tetrahydrogestrinone ("thg" or "the clear"). Taken as pills or injections, they encourage an increase in muscle bulk to improve strength, speed, and recovery from injury. They are most commonly abused in brief-action "power" sports such as sprinting, weightlifting, and hammer and other throwing events.

PAINKILLERS

Effects: Allow athletes to compete in spite of pain or injury.
Main risks: Severe and permanent damage to body parts, shock, collapse.

• Animal testing of new medicines is required by law in most countries.

• The AIDS microbe HIV was originally tested on chimpanzees, but their bodies were found to react very differently to those of humans.

• People who volunteer to receive new medicines are usually paid, but never too much, to discourage people continually offering themselves.

• Genetic researchers create strains of rats, mice, and other laboratory animals with a ready-made tendency to develop diseases such as cancers.

Animal Testing of Medicines

When you take a medicine, health supplement, or food additive, apply a cosmetic, or even put toothpaste on your brush, of course you want to know that the product is safe. How is this done? By testing such substances in a series of ways—first using laboratory equipment, then on cells cultured (grown) in flasks and dishes, next on animals, and finally on human volunteers. The live animals used in such testing include mice, rats, guinea pigs, rabbits, dogs, cats, and monkeys. The tests show whether the substance has the intended benefits, and whether it has additional harmful or toxic effects. Worldwide, hundreds of thousands of animals die every year as a result. But in almost all countries, such tests are demanded by law before a substance can be given to humans.

Animal testing should not be done because...

Apart from various religious, moral, and ethical arguments not based on science:

• Animals aren't good enough biological "models" for the human body. The response of an animal to a substance does not mean a person will react in the same way.

• Stress experienced by animals can affect the test results.

• Alternative testing methods are available (see panel, right).

• It's expensive. The cost of testing a single medicine thoroughly on animals is frequently several million dollars.

• Usually it's a waste. Very few possible drugs and other substances get past the animal test stage.

• Animal welfare laws have not stopped some animals suffering pain and cruelty.

NO

X

STOP ANIMAL CRUELTY NOW!
Some people feel very strongly about animal testing, but almost all of them will, at some stage in life, rely on medicines that have been tested on animals.

Looking Forward to the Future

Scientists who develop new medicines, cosmetics, and other products are far from heartless, unthinking robots. Many are continually torn between the need to make their products as safe as possible for people, and the necessity of testing them on animal "guinea pigs." They are always trying to refine animal tests, reduce the animals' possible suffering or distress, and improve their welfare.

New non-animal methods are constantly being explored. This is partly to save animals, but also to allow manufacturers to use advertising phrases like "Not tested on animals." Despite all these ongoing advances, if we want new, better, safer medicines, and other products, then some testing on live animals is likely for several years to come. A ban now on all animal testing would probably mean no new medicines or similar products for 20–30 years at least.

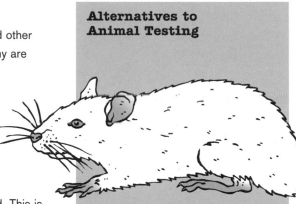

Alternatives to Animal Testing

Scientists are increasingly working on new methods that cut down on the amount of animal testing needed, including:

• Laboratory cultures of single-celled life-forms, such as bacteria, tested for cancer-causing potential.

• Cultures of animal or human cells and tissues. For example, substances can be checked for their potential to cause skin irritation on artificial human skin.

• Microdosing, where substances are checked out in extremely tiny amounts in human volunteers, and increased very, very slowly.

• Computer simulations that predict what effect a substance may have on a body process or system.

Animal testing is needed because ...

• Nearly all medical advances over the past century have relied on animal testing, from vaccines to life-saving drugs for common problems like asthma, diabetes, and heart disease. Without these advances, we would be much less healthy than we are today.

• Testing medicines and other substances on living animals is often the only way of finding out many important aspects of how they work.

• Trying to replace such tests—for example, with computer simulations or cells in dishes—cannot provide as much reliable and realistic information as a whole living system can give.

• In most countries where testing is done, animal welfare laws protect against unnecessary cruelty as much as possible.

8

Genes Stuff

CHAPTER CONTENTS

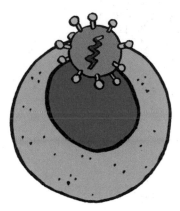

The science of genetics has advanced at a tremendous pace over the past half-century. The structure of deoxyribonucleic acid (DNA), which forms the physical basis of genes, was discovered only in 1953. It was another 13 years before the rudiments of how DNA works had been untangled. By the mid-1970s, however, scientists had discovered how to extract genes from the cells of one organism and insert them into another for the purpose of making medically important drugs. Techniques of this type came to be known as "genetic engineering" and were soon extended to modifying the genetic material of plants, creating GM (genetically modified) crops.

Human Genetics

Starting in 1990, scientists across the world joined in a major project to identify and map all human genes. This became known as the Human Genome Project (HGP) and by 2003 it was complete. One of its findings was that there is relatively little genetic variation among humans. For example, a comparison of the DNA of two people randomly chosen from the world population, will reveal that their genes are around 99.5 percent identical. Since the completion of the genome project, scientists have focused on working out what role each gene plays in the body and what diseases are caused when particular genes contain errors. This painstaking work—a vast undertaking of colossal complexity—has already discovered faulty genes associated with a variety of diseases. Knowing about such genes helps identify people who might be prone to the relevant diseases and also assists in devising appropriate treatments.

Genes and Genetics Explained

The chapter starts with a brief overview of what genes and DNA are, and how they function, followed by a description of how genes are passed on from parents to offspring, and how they interact. We look briefly at the branch of genetics that has provided clues to the historical origins and spread of modern humans; other sections deal with gene therapy, genetically modified foods, and DNA fingerprinting, which over the past 20 years has become an invaluable crime-solving tool. Toward the end of this chapter are sections on the controversial topics of cloning and stem-cell research.

- Genes are units of hereditary material, contained within the nuclei of body cells. They hold the instructions for how your body works.

- Each person has about 30,000 genes. These are divided between 23 pairs of structures called chromosomes in cell nuclei.

- Chromosomes consist mainly of the complex thread-like chemical substance called DNA. Genes are made up of lengths of DNA.

- There is not very much variation between different people's genes or DNA—genetically, humans are basically all very similar.

What are Genes?

Everyone holds within each of their body cells a copy of what is called human DNA or the human genome. This DNA (deoxyribonucleic acid) contains a person's genes, the coded instructions for the body's physiological characteristics and how it functions. Everyone's DNA or set of genes—the terms are more or less interchangeable—is almost identical, but there are still quite a few differences between one person and the next (identical twins excepted), and it is these differences that determine each person's unique individual characteristics, including their eye color, height, and many other traits including susceptibility to some diseases.

4. Each chapter consists of many sentences, which contain the instructions for aspects of how your body works or is built. We call these sentences "genes." In most cases, the sentences in a chapter provided by your father are exactly the same as the sentences in the equivalent chapter supplied by your mother. For example, two equivalent sentences in paired chapters concerned with facial anatomy might both specify "Number of eyes: Two." But in a few cases there are differences. For example a sentence that has come from your mother might say "Color of eyes: Brown," while the equivalent sentence from your father might say "Color of eyes: Blue."

3. The instruction manual contains 46 chapters, but these can be thought of as 23 pairs of chapters. In each case, one of the pair has come from your mother and the other from your father.

DNA in cell nuclei

1. Your DNA is like an instruction manual, containing all the information on how your body should be put together and how it works.

Body cells

2. A copy of this DNA manual is contained in the nucleus of every body cell (a small amount of DNA is held in structures outside the nucleus called mitochondria).

CHAPTER 5a

CHAPTER 5b

Chapter (chromosome) from mother

Equivalent chapter (chromosome) from father

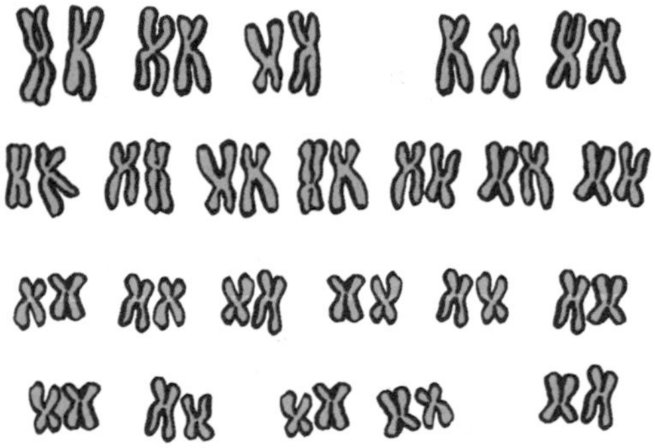

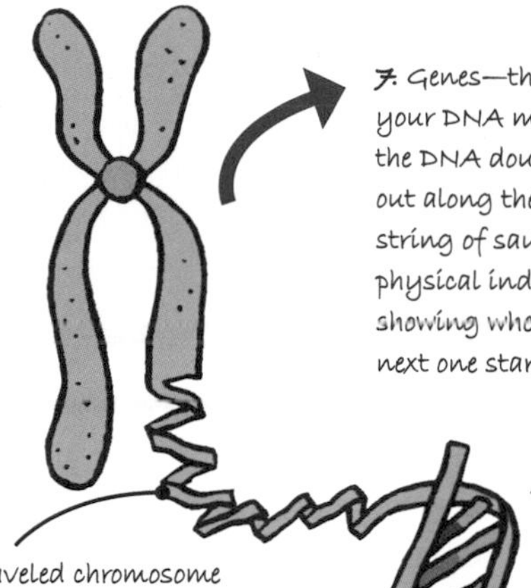

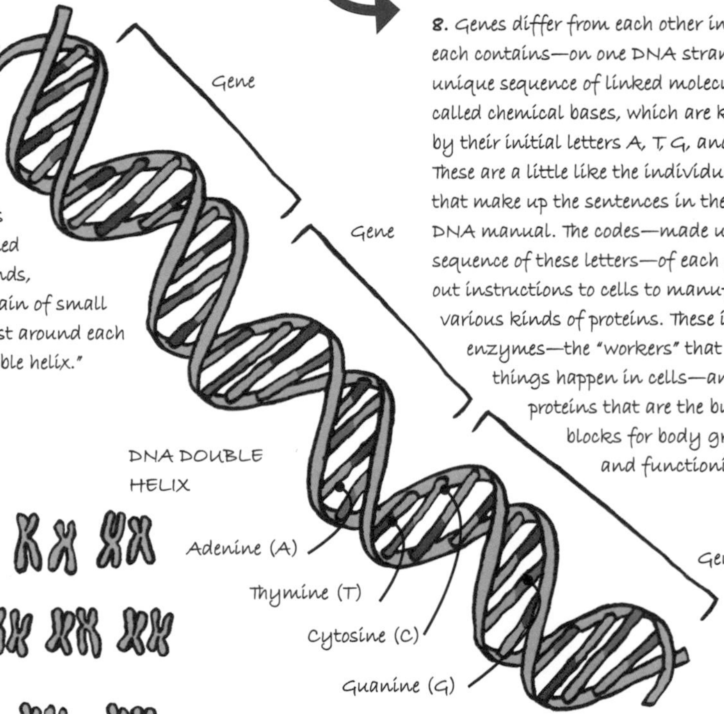

7. Genes—the sentences in each chapter of your DNA manual—are various parts of the DNA double helix. The genes are strung out along the DNA thread. But unlike a string of sausages, there are no obvious physical indicators on a DNA thread showing where one gene ends and the next one starts.

GENOME: The complete set of genes, or DNA, of an organism, whether a human or other animal, a plant, or other living thing.

unraveled chromosome

6. Each chromosome consists principally of a long, tightly coiled thread of DNA. If the DNA is unraveled, it can be seen to be formed from two chemical threads, or strands, each of which consists of a long chain of small molecules linked together, that twist around each other. This is the famous DNA "double helix."

8. Genes differ from each other in that each contains—on one DNA strand—a unique sequence of linked molecules called chemical bases, which are known by their initial letters A, T, G, and C. These are a little like the individual letters that make up the sentences in the human DNA manual. The codes—made up of a sequence of these letters—of each gene send out instructions to cells to manufacture various kinds of proteins. These include enzymes—the "workers" that make things happen in cells—and other proteins that are the building blocks for body growth and functioning.

Gene

Gene

DNA DOUBLE HELIX

Adenine (A)

Thymine (T)

Cytosine (C)

Guanine (G)

Gene

Chromosomes

5. At the level of cells and nuclei, the 23 pairs of chapters in the manual represent the 23 pairs of structures called chromosomes (46 chromosomes in all) in your cell nuclei. All the DNA in your cell nuclei is contained within these chromosomes. Chromosomes are typically depicted as having an X-shape, as shown here, although they look like this only for a short time—just before cells divide.

Sex Chromosomes

Of the 23 pairs of chromosomes, one pair, called the sex chromosomes, are different from the rest in an important respect. In females, these two chromosomes are a fully equivalent pair and are called X chromosomes: They contain genes for general body functioning. In males, however the two sex chromosomes aren't really an equivalent pair at all. One is an X chromosome, as in females, but the other is a special chromosome found only in males, called the Y chromosome. This small chromosome contains genes that produce or code male characteristics. A male always receives his Y chromosome from his father, while his X chromosome comes from his mother.

- Unrelated humans are typically 99.5 percent genetically identical. The remaining 0.5 percent of genes account for the differences between them.

- We inherit our genes equally from both parents. Identical twins excepted, offspring of the same parents always receive a different "mix" of their parents' genes.

- Many traits, such as eye color, are determined by one or more gene pairs, each member of each pair having come from a different parent.

- Where the genes in a pair are different, they interact according to specific rules to determine how a person is affected.

Genes and Inheritance

You inherit your genes from your parents—half from your mother, via one of her egg cells, and half from your father, via one of his sperm cells. In fact, most of the genes that a person possesses exist in pairs, which are held on paired chromosomes, the two members of a pair having come from that person's two parents. Often the genes in a pair are identical. But for genes involved in determining characteristics that vary a lot between people (such as eye color and blood type), the two members of a pair are sometimes different. Where this happens, they interact in specific ways to determine how a person's appearance or functioning is affected.

Gene Interactions in Action

To understand how genes work together to determine inherited characteristics, consider the example of the fictional matchstick people (see far right). Where two genes in a pair are different, typically one masks, or is "dominant" over, the other (see *Gene Interactions in Matchstick People*, right). In real people, the same sorts of rules apply, although they are more complex. Because of the random way in which the two members of each gene pair are allocated to a matchstick person's (or real person's) egg or sperm cells and thus to his or her children, the offspring of a particular couple always receive a different "mix" of their parents' genes—identical twins being the only exception. Depending on the components of this mix, offspring often resemble one parent or a sibling in some characteristics and the other parent or a different sibling in other characteristics, but because of the way in which gene pairs interact, they may also display traits that are absent from all other family members.

Gene Interactions in Matchstick People

If the two genes a matchstick person possesses for a particular characteristic are the same, the outward appearance is simply as specified, for example:

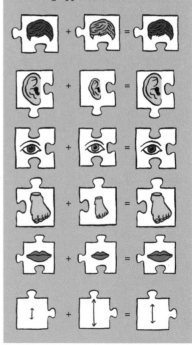

If the two genes are different, the following applies:

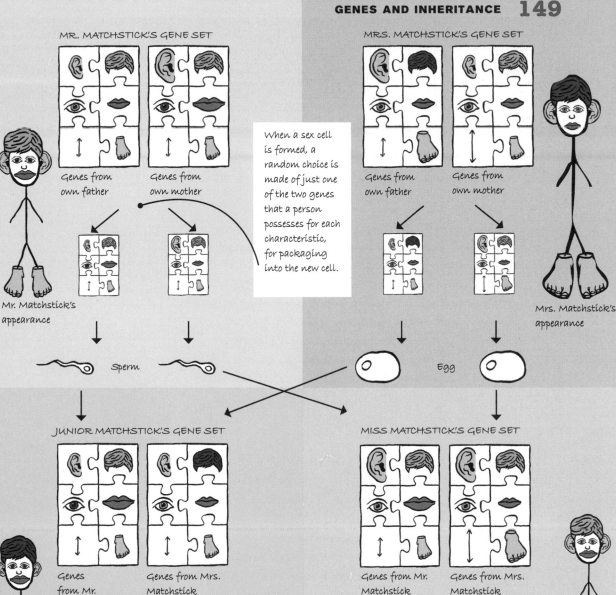

MR. MATCHSTICK'S GENE SET

Genes from own father

Genes from own mother

When a sex cell is formed, a random choice is made of just one of the two genes that a person possesses for each characteristic, for packaging into the new cell.

MRS. MATCHSTICK'S GENE SET

Genes from own father

Genes from own mother

Mr. Matchstick's appearance

Mrs. Matchstick's appearance

Sperm

Egg

JUNIOR MATCHSTICK'S GENE SET

Genes from Mr. Matchstick

Genes from Mrs. Matchstick

MISS MATCHSTICK'S GENE SET

Genes from Mr. Matchstick

Genes from Mrs. Matchstick

Junior Matchstick's appearance (like father but with smaller ears and hair like mother).

Miss Matchstick's appearance (like mother but with hair and mouth like father's).

Genetic Inheritance in a Matchstick Family

Matchstick people have a much simpler gene set than humans, with just six characteristics determined by genes that vary from one person to another. These are hair shade, ear size, eye color, mouth size, foot size, and height. Each of the six characteristics is controlled by a pair of genes, one of which has come from the matchstick person's father and the other from the matchstick person's mother. The genes interact following the rules in the panel on the left.

• All people alive today are descended from a much smaller pool of people who lived in the distant past.

• Some bits of human DNA, such as something called mitochondrial DNA, can be analyzed to reveal what mutations it has undergone over many generations.

• Analysis of mitochondrial DNA can reveal how different population groups in the world today are related and when they diverged.

• All living people have had their mitochondrial DNA passed down to them from one woman who lived tens of thousands of years ago.

Genetic Clues to Human Past

Humans don't vary enormously in their genetic make-up. What differences there are have resulted from mutations to human DNA over many generations, during the evolution of today's human population from a much smaller pool of common ancestors who lived long ago. Because our genes or DNA occur in new combinations with each generation (see pages 148–149), ordinarily it would be impossibly complex to work out when, and in what order, different mutations have occurred over the generations. But since the 1970s, scientists have discovered that parts of human DNA can in fact be analyzed in this way—and their investigations have provided fascinating insights into the human past.

Mitochondrial DNA

A small amount of everyone's DNA is held outside the cell nucleus in structures called mitochondria. An important characteristic of this mitochondrial DNA (mtDNA) is that people always inherit it from their mothers. Every living person can trace their mitochondrial DNA back through their female ancestors to one woman, who is estimated to have lived between 160,000 and 220,000 years ago and has been named "Mitochondrial Eve." Studies on mtDNA, and also on another piece of DNA called the Y chromosome, which is passed down only from fathers to sons, have provided a vast amount of information about the human past.

Mitochondrial Eve's "Heirloom"

Mitochondrial DNA, or mtDNA, is like an heirloom, a copy of which is passed to every person from their mother. In this way, copies of the heirloom have descended down from Mitochondrial Eve, through a succession of mothers and daughters, to everyone alive today. On the way the mtDNA has acquired mutations (changes), which are like decorations added to specific copies of the heirloom. As a result, all people alive today divide into distinct groups that are characterized by particular patterns of mutation in their mtDNA, or patterns of markings on their "heirlooms."

What Does mtDNA Reveal?

Because mtDNA has been passed down to everyone alive today only through females, without getting mixed up with genetic material passed through males, it is relatively easy for geneticists to analyze the mutations that have occurred to it as it has traveled through time.

The comparison of samples of mtDNA from people across the world and a study of the pattern of mutations enables geneticists to work out the order in which the mutations occurred. What's more, because mtDNA mutates at a fast and regular rate, it is also possible to estimate roughly when these mutations occurred and so provides an estimate of how long ago Mitochondrial Eve lived.

One of the main findings of this research is that people with particular patterns in their mtDNA tend to be concentrated in particular regions of the world, making it possible to work out how different population groups are related to each other, as well as roughly how long ago, and even where, they diverged as a result of population "splits" and migrations. For example, the most ancient divergences occurred in Africa over 100,000 years ago, with the implication that modern humans originated in Africa. About 65,000 years ago, somewhere in Western Asia, there was a major split among a band of people who had left Africa, with one group going on to populate much of East Asia, Australasia, the Pacific, and eventually the Americas, while the other spread to the rest of Western Asia and all parts of Europe.

Can We be Sure that She Existed?

Simple logic shows that if everyone traces their ancestry back through the female line only, these ancestral lines eventually converge on one woman. Everyone living today has one, and only one, biological mother, but many have the same mother, so there must be fewer mothers (living or dead) of people alive today than the number of people alive today. Going back one generation, some of those mothers themselves have the same mother, so there must be fewer maternal grandmothers (again, living or dead) of people alive today than mothers of people alive today. Maternal great-grandmothers are even fewer in number than maternal grandmothers, and so on. Thus with each generation going back, the number of women who are ancestors through the female line of living people gradually diminishes, until eventually it arrives at one—the person we know as Mitochondrial Eve.

MUTATION:
A change in a gene that is the result of an error in copying the DNA in a cell during the process of division into daughter cells.

Jargon buster

The Paths Back to Mitochondrial Eve

If the maternal ancestry lines of everyone alive today were extended back, they would eventually converge on one woman who lived thousands of generations ago and has been called Mitochondrial Eve. She was not the only woman alive at her time but just one of a group of men and women.

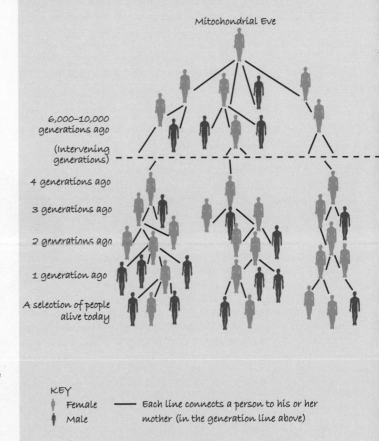

Mitochondrial Eve

6,000–10,000 generations ago

(Intervening generations)

4 generations ago

3 generations ago

2 generations ago

1 generation ago

A selection of people alive today

KEY
Female
Male
—— Each line connects a person to his or her mother (in the generation line above)

• Mutations (changes) that have occurred in the past to genes are the cause of a number of mostly rare disorders.

• Gene therapy most often refers to "somatic gene therapy," an experimental and, so far, uncommonly used approach to treating some genetic disorders.

• The most common approaches to the management of genetic disorders include drug treatments and other therapies, as well as the provision of counseling on the risks of passing on a disorder to future generations.

Gene Therapy

A mutation, or change, in human DNA can sometimes result in the creation of a faulty gene that is unable to carry out its function in the body. This can result in various degrees of abnormality in affected individuals. Often, the basic problem is that the affected person's body lacks a particular enzyme, other protein, or hormone, which in most people is made by a normal version of the same gene.

One experimental approach to treating these disorders, which may become more important in future, is called somatic gene therapy (see opposite page). This approach has already been used with some success to treat a number of different genetic disorders.

Germline Gene Therapy

This is a radical and controversial proposed approach to treating genetic disorders that has not yet been used. It would involve trying to correct or replace a defective gene in a germline cell, such as an egg cell. Theoretically, this could prevent the disorder from affecting any of a person's descendants, but there are complex ethical issues surrounding such an approach as well as technical obstacles.

Conventional Approaches to Genetic Disorders

Most genetic disorders cannot, at present, be treated by somatic gene therapy, although some can be treated by more conventional treatments with, for example, drugs that correct the biochemical deficiency caused by the disorder. Another important approach to seriously debilitating genetic disorders is genetic counseling. This involves providing advice on the risks of passing on a genetic disorder to offspring and the options to consider to minimize this risk. Information about genetic disorders gathered as a result of the Human Genome Project and other scientific studies (see page 145) has enhanced the advice and treatment that can now be given to sufferers from genetic disorders.

Somatic Gene Therapy in Action

This form of treatment involves introducing a normal gene into the affected person's body, most often by first inserting it in a lab into a sample of the person's cells.

1. Some somatic cells (see Jargon buster, below) are removed from the person to be treated.

2. In a laboratory, a virus is modified so that it is no longer able to cause illness.

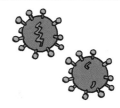

3. A normal version of a particular gene—the gene that the person being treated lacks —is inserted into this modified virus.

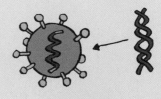

4. The altered virus is inserted into the cells from the person to be treated.

5. In this way, the cells become genetically altered, as they incorporate the beneficial normal gene.

6. The genetically altered cells are injected back into the person to be treated. There they begin to produce the desired protein or hormone that the person is lacking.

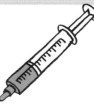

Jargon buster

SOMATIC VERSUS GERMLINE: A somatic cell is a general body cell, such as a skin, nerve, or bone cell. In contrast, a germline cell is an egg or sperm or fertilized egg. Genetic changes to germline cells can affect subsequent generations of people, whereas a genetic change to a somatic cell affects only the person in which that cell resides.

• Almost 425,000 square miles of Earth's land surface—about 2.4 percent of all agricultural land— is now devoted to growing GM crops.

• About three-quarters of all processed foods in the US contains at least one GM ingredient.

• Concerns about GM foods focus mainly on possible damaging effects on the world's ecosystems.

• At present, there is no evidence that eating GM foods is damaging to health, although possible long-term adverse effects can't be ruled out.

GM Foods

GM (genetically modified) foods are most commonly plants that have had a gene or genes added or altered to change their characteristics in some way, usually to make them easier and cheaper to grow, to make them resistant to pests such as insects, to increase their nutritional content, or to make them stay edible for longer after they have been harvested. In theory, GM foods reduce production costs and increase yield, so food becomes less expensive and more abundant. But in practice, there is little evidence that GM crops have increased yields beyond what has been achieved using traditional plant breeding in previous decades.

Cost-Savings Can Be Short-Lived

One way in which certain GM crops—those that resist being eaten by some insects—are claimed to reduce farming costs is that less insecticide has to be purchased. This is often true initially, but as the original insects die off, other species may thrive in their place and start attacking the crops. The farmer then has to increase insecticide use again, which can negate the original benefits.

Typical genetically modified food crops include maize, soybeans, and rice. Animal products have also been developed. So far, more than 20 countries have embraced the production of GM foods, including the US, Canada, Brazil, Argentina, and China. But many others have yet to accept GM foods or have banned their production, due to concerns about possible adverse effects on human health or the environment.

How Plants are Genetically Modified

To genetically modify a plant, first a gene with a useful characteristic has to be identified and isolated. Typically, this gene is then inserted into a piece of DNA called a plasmid, which is relatively easily transferred into the nucleus of a cell from the plant to be modified. Subsequently, the plasmid incorporates itself into one of the cell's chromosomes. The modified cell is then encouraged to start dividing, and soon enough one has a genetically modified plant.

Weedkiller-Resistant Crops

A common way in which plants are genetically modified is to make them resistant to a herbicide (weedkiller), so that farmers can easily kill weeds on the land where the GM plant crops are grown without affecting the food plant. This has undoubtedly encouraged farmers who switch to the relevant GM crops to greatly increase use of herbicides. This is a matter of concern to environmentalists as some studies have shown these herbicides to be toxic to wildlife.

Objections to GM Plants and Foods

A number of objections and concerns have been raised about genetic modification of food crops, including:

• If genes are spread from modified crops to wild relatives, those relatives could be changed in a way that could make them play a different ecological role, potentially enabling them to outcompete other species, with consequent adverse effects on plant biodiversity.

• There are concerns about the ecological effects of replacing normal crops with GM crops, together with accompanying changes in herbicide use. For example, increased use of herbicides may encourage the evolution of species of weeds resistant to herbicides. Conversely, wildlife species that are reliant on the weeds may be wiped out.

• There are fears that GM foods may contain unidentified toxins that have long-term adverse health effects. So far, no such toxins have been found, and these fears appear to be unfounded. However, some scientists think that current testing methods are not rigorous enough.

Jargon buster

BIODIVERSITY: The degree of diversity of different life forms, regarded as an indication of the ecological health of a region.

- Cloning can refer to the copying of DNA fragments, cells, or whole organisms.

- There is a distinction between therapeutic cloning—the production of copies of a particular type of cell for the purpose of treating disease—and whole animal cloning.

- Individual animals from about 20 different mammal species have been cloned as of 2009.

- Attempts at whole-animal cloning frequently lead to embryos or live-born animals with severe defects, and a similar outcome is likely if attempts are made to clone people.

Cloning

Cloning is the production of identical copies of something. In nature, many organisms such as bacteria, fungi, and plants reproduce by making copies of themselves. In biotechnology, cloning can refer to the copying of DNA fragments (molecular cloning), cells, or whole organisms (reproductive cloning). Therapeutic cloning refers to the copying of a class of human cells, called stem cells, in order to develop new treatments for disease (see page 160–161).

Jargon buster

BIOTECHNOLOGY: The use of living organisms, or parts of living organisms, for any of various industrial, agricultural, or medical applications. It includes genetic engineering and cloning methods.

Cloning Endangered Species

A possible reason suggested for cloning animals has been to help save endangered species or even to resurrect extinct species. In January 2009, a Pyrenean Ibex (type of goat) was the first animal belonging to an extinct species to be born. This was achieved using cloning techniques on frozen cells taken from an ibex that had died many years earlier.

In the popular media, the word cloning is most often used to refer to the cloning of whole organisms by a technique called nuclear transfer (see *Cloning a Cat*, opposite). This has been done, for example, in sheep (initially with a ewe named Dolly), mice, cats, a camel, and other animals. The technique could also theoretically be used to clone humans (there are unverified claims that it already has been used in this way). A human produced by this technology would not be genetically identical to the original, because not all human DNA is contained in the nucleus (see pages 146–147). A clone of a human would also certainly not be "the same person" as the original but a unique individual, in the same way that identical twins are different individuals.

Risks of Cloning

Attempts at cloning animals often go wrong. According to a 2002 survey, about a quarter of all cloned mammals produced by nuclear transfer fail to reach healthy adulthood. If attempts were made to reproductively clone humans, it is envisaged that a lot of babies would similarly be born with defects of varying degrees of severity.

8. The kitten born was a clone of the cat from whom body cells were taken.

Cloning a Cat

The first cloned cat—which came to be known as Copy Cat—was produced using the method of nuclear transfer shown here. Since then, a slightly different technique, called chromatin transfer, has been used to clone cats. In chromatin transfer, only the chromosomes, rather than the whole nucleus, of a cell from the cat to be cloned is transferred into a donor egg.

1. A female cat acted as an egg donor.

Unfertilized donor egg

2. In the laboratory, the nucleus was removed from the egg.

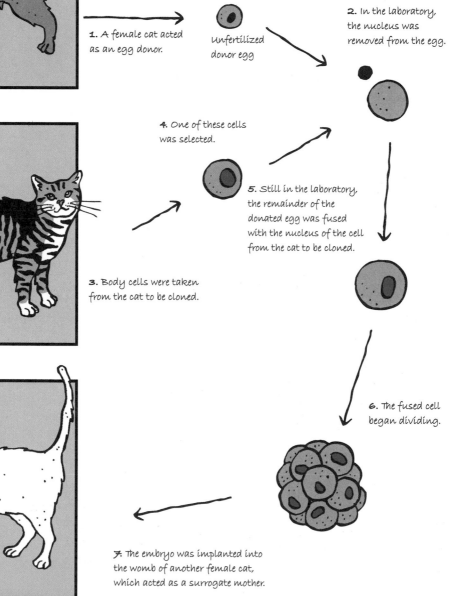

4. One of these cells was selected.

5. Still in the laboratory, the remainder of the donated egg was fused with the nucleus of the cell from the cat to be cloned.

3. Body cells were taken from the cat to be cloned.

6. The fused cell began dividing.

7. The embryo was implanted into the womb of another female cat, which acted as a surrogate mother.

- DNA fingerprinting is a technique that has proved particularly useful in solving crimes and in paternity testing.

- DNA fingerprinting can often rule out a suspect from a crime, or a man from being the father of a specific child.

- DNA fingerprinting is distinct from full genome sequencing, which is the complete recording of a person's DNA.

- The chance of a coincidental match in the DNA fingerprints of two individuals (as long as they are not identical twins) is about one in 100 billion.

DNA Fingerprinting

DNA fingerprinting is a scientific technique used to help identify individuals in criminal and other cases. In the 1980s, it was shown that certain sequences of the chemical bases that make up DNA vary significantly across the population, and that the overall pattern of these sequences is unique to each person (identical twins excepted). A single cell from a biological sample

Crime-Solving Using DNA Fingerprinting

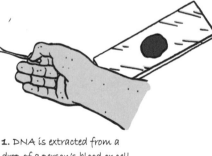

1. DNA is extracted from a drop of a person's blood or cell sample or from biological material (such as blood, saliva, or hair) found at a crime scene.

2. The DNA is cut up by the action of proteins called enzymes. Because each person's DNA is slightly different, this results in a unique collection of DNA fragments.

DNA MARKER: A DNA sequence that has a known location on a chromosome and varies between individuals. A DNA fingerprint is essentially a collection of such markers.

3. Using an electric field, the fragments are arranged by length on the surface of a gel, and are then transferred to a nylon membrane.

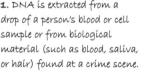

4. Radioactive chemicals are used to tag certain "marker" fragments, and the membrane is then exposed to X-ray film. When developed, this displays a pattern of banding, the DNA "fingerprint."

Tagged DNA marker

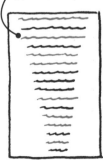

DNA Databases

The effectiveness of DNA fingerprinting in solving some crimes depends on police holding a database of many peoples' DNA fingerprints, with the fingerprints of people who have previously committed serious crimes being particularly desirable database components. Many people believe, however, that DNA databases infringe civil liberties when they contain the DNA fingerprints of people who have never been convicted of any crime or have committed only minor offenses. In relation to this, the European Court ruled in 2008 that keeping the DNA records of people never convicted of a crime breaches an article of the Human Rights Convention and that such records should be destroyed.

Paternity Testing

In addition to its use in solving crimes, DNA fingerprinting can be used to help resolve paternity disputes. All of a person's DNA markers should be identifiable either in that person's mother or in the biological father. Here, of the child's markers that aren't present in the mother's DNA—fragments A, B, and C—each is present in Matt's DNA fingerprint but not in Tom's. This means that Matt could be the biological father but Tom could not. In practice, DNA fingerprints are longer than shown here, and to prove beyond all reasonable doubt that Matt is the father would require the identification of several more matches between the child's markers (after subtracting those attributable to the

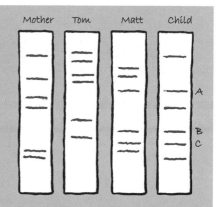

mother) and Matt's. It might also need to be established that Matt does not have an identical twin brother.

is enough for laboratory analysis of that sample's DNA, producing a "fingerprint," similar to a bar code, of the person who left the sample. In practice, the sample may be provided by blood, cells taken from the mouth, saliva, semen, or cells taken from personal items such as razors and toothbrushes, depending on the circumstances.

DNA fingerprinting is highly accurate and reliable. The statistical chances of fingerprinting producing a misidentification are generally extremely low, although the existence of an identical twin can occasionally be a complicating factor that has to be taken into account in legal proceedings.

5. In this example, the DNA pattern of a sample found at the scene of a crime exactly matches the DNA of one of three suspects, leading to an immediate arrest.

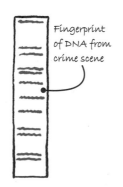

Fingerprint of DNA from crime scene

Matching fingerprint

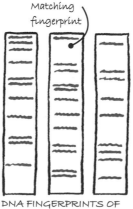

DNA FINGERPRINTS OF THREE SUSPECTS

- Stem cells are types of cells that have the potential to divide and develop into many different types of cells and tissues.

- Embryonic stem cells, taken from developing embryos, are the type of stem cell with the most potential for developing into different cell and tissue types.

- Research on embryonic stem cells is aimed at producing various types of medical advance, including treatments for disease.

- Arguments about whether embryonic stem-cell research should be allowed are based on differences in belief about whether an embryo is or is not a person.

Stem-Cell Research

In many parts of the world, couples with difficulties having children may be helped by a technique called in-vitro fertilization (IVF). In this procedure, eggs taken from a woman are fertilized in a laboratory using sperm from her partner or another donor. Some of the healthy embryos produced by the process are implanted in the woman's womb. The rest are usually frozen so that they can be used in future should a pregnancy not develop. Potentially, however, some of these embryos may not be required by the couple and may be used by scientists for research purposes. One particular type of research involves cells taken from embryos called embryonic stem cells. This inevitably leads to the destruction of the embryo. Embryonic stem cells are especially valuable for scientific research because they can differentiate into all of the 220 types of cells found in the human body and have huge potential for treatment of various diseases. (Differentiation is the complex process by which an unspecialized cell develops through a series of divisions into a specialized cell, such as a nerve cell.)

People are divided on whether research on embryonic stem cells should be allowed. It may be possible to sidestep this controversy in the future, as techniques are being developed for converting human skin cells into cells that mimic embryonic stem cells, and might offer their advantages without the ethical problems.

Jargon buster

IN-VITRO: Literally means "within a glass." In vitro procedures are techniques used for manipulating biological materials in a laboratory—classically in glass test tubes.

The Ethical Debate

Arguments For:

Embryo and embryonic stem-cell research have the potential to produce various types of medical advances, such as new prenatal tests for detecting genetic diseases and therapies for the chronic diseases of old age. Scientists might be able to grow a particular type of cell in the laboratory and then inject it into a patient, where it would replace diseased tissue. But stem cells are not yet being used for this purpose because scientists haven't yet learned how to direct them to differentiate into specific tissues or cell types. Supporters of the research needed to achieve these aims generally believe that while an embryo has the potential to develop into a person, it is not yet a person, and that the research is therefore ethical and justifiable.

Fertilized human egg

Potential Uses of Stem Cells
Stem cells taken from embryos, as shown here, have a particularly high potential for developing into many different types of cell or tissue, which could be used to treat diseases.

Adult Stem Cells
In addition to embryonic stem cells, there are other types of stem cell called adult stem cells. Research on these is much less controversial, but adult stem cells are potentially less useful medically than embryonic cells.

Egg after first few divisions

Therapeutic Cloning
Therapeutic cloning involves the creation of an embryo-like cluster of cells that originate not from the fertilization of an egg by a sperm, but by the replacement of the nucleus in a donated egg with the nucleus taken from a cell of the person for whom the treatment is intended. The aim is to produce stem cells that are genetically matched to that person. These might be used in the future to treat any of various chronic diseases in that person without the problem of those cells being rejected by the person's immune system.

Early embryo

Arguments Against:
Objections to embryo and embryonic stem-cell research are mainly based on a belief that human life starts at the time of fertilization, rather than at the point when the embryo implants in the mother's womb, or at a later time. Adherents to this view believe that an embryo is already a person with rights, and that any research on embryos is therefore wrong. Many adherents to the same view also object to specific aims of embryo research, such as the development of improved prenatal tests, on the grounds that these imply the possibility of aborting fetuses that are carrying genetic defects or other abnormalities.

Embryonic stem cells

Cultured embryonic stem cells can potentially be grown into many different types of body cell.

Heart-muscle cells

Blood cells

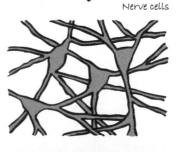
Nerve cells

9

Harder Stuff

This chapter focuses on some theories in physics that are sometimes thought of as hard to understand. While some of their details are indeed complex, the gist of what they say is less difficult to grasp. These theories possess an inherent fascination because some of their predictions run counter to intuition. The special theory of relativity, for example, maintains that time can pass at a variable rate; quantum theory says that particles are sometimes waves; while one "theory of everything" maintains that all matter is composed of exceedingly tiny strings.

A Failed Experiment

One of these theories, special relativity, has its beginnings in an experiment performed in 1887, which has been called the most famous failed experiment in the history of science. At that time, scientists assumed that the motions of all objects could be measured relative to an absolute, universal, "frame of reference" (think of a 3D grid filling the whole of space). This was thought to contain an invisible material called "the ether," which could carry light. The 1887 experiment aimed to measure how fast Earth was moving through

this ether by sending light rays in different directions, bouncing them off mirrors, and measuring how long it took them to return—it was assumed that a "wind" created by Earth's movement through the ether would slow the light when it was traveling in certain directions. The experiment failed dismally because whatever direction light was sent in, it always traveled at the same speed. The scientists were baffled, but at no point questioned either the idea of the ether or the concept of an absolute reference frame.

A Scientific Revolution

It was not until 1905 that a fully worked-out explanation was proposed, by a young German-born physicist called Albert Einstein. Einstein took it as a "given" that light always travels at a constant speed. His theory rejected many deeply ingrained notions— including the idea of the ether, the concept of an absolute reference frame, and ultimately the idea that time passes at the same rate everywhere. The theory shows how scientific progress sometimes occurs by revolution—through the radical overturning of established ideas—rather than by gradual evolution.

bite size facts

• Special relativity is based on the principle that the speed of light is constant, whatever the relative movement of the light source and an observer.

• Special relativity also predicts that from the point of view of an external observer, time slows down for an object moving near the speed of light.

• In relativity theory, the three dimensions of space and one of time are replaced by a four-dimensional system, called space-time.

• The more complex "general" theory of relativity proposes that gravity produces its effects as a result of a mass distorting space-time.

Einstein's Relativity

There are two separate theories of relativity in physics: special relativity and general relativity. These were both developed by the Swiss–German physicist Albert Einstein in the early part of the 20th century.

These theories deal with various matters concerned with motion, time, mass, energy, and (in the case of general relativity) gravity, that previously had been described by what is called "classical mechanics." To this day, classical mechanics remains valid for describing the everyday, familiar world. But for certain more extreme situations, such as particles moving at exceedingly high speeds or the properties of black holes, the theories of relativity provide a more precise framework. Though initially regarded with some skepticism, the predictions of both special and general relativity are now known to be highly accurate. The understanding of relativity also has practical applications—GPS (global positioning system) technology, for example, works only if the effects of relativity are taken into account.

Special Relativity

Special relativity was the first and simpler of Einstein's two theories of relativity. Einstein based it on two principles. The first is that the speed of light is always the same, at about 300 million meters per second, whatever the relative movement of the observer and light source. The second is that the laws of physics work in the same way for all objects or systems of objects, whatever their relative movements. By the logical application of these principles, Einstein was able to demonstrate various startling results, which seemed to run counter to human intuition. For example, he was able to show that there is no such thing as two simultaneous events, because two events recorded as happening simultaneously by one person will seem to occur at different times when observed by a second person moving relative to the first.

Einstein also showed that when an object is traveling at very high speeds, from the point of view of an external observer, it shortens in length and the passage of time slow downs for that object, even though for someone who is actually part of the object, time seems to pass normally. Furthermore, when traveling at close to the speed of light, the increase in an object's energy has an effect like that of an increase in its mass, and it becomes increasingly difficult to accelerate the object any further. This led to the perception that mass and energy are actually two different manifestations of the same thing, called mass-energy, and to the famous formula $e=mc^2$ (see page 34).

1. A tourist bus, 30 feet (10 m) long, is parked next to a high-speed railroad. The tourists have come to see the "Speed of Light" train. To check on the train's speed, two cameras connected to highly sensitive stopwatches are mounted on top of the bus—one at each end.

2. The train zooms by. As its leading edge passes first the rear of the bus, then the front, these events are recorded by the cameras and stopwatches. The stopwatches show that there was a time difference of one thirtieth of a millionth of a second, indicating that the train was traveling 300 million meters per second—the speed of light.

3. The bus then sets off and is driven very fast along the highway, beside the railroad, getting up to a speed of about 90 percent of the speed of light.

4. Another Speed of Light train goes by. Once again, the time for its leading edge to pass from the back of the bus to the front is recorded by the cameras and stopwatches. The bus itself was traveling so fast, the tourists thought the train would take longer to overtake them this time.

5. They were wrong! The time difference is still the same—the train still moved relative to the bus at 300 million meters per second, the speed of light.

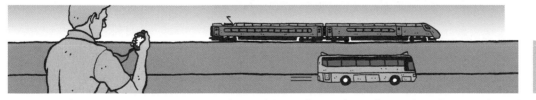

6. On a nearby hill, a train enthusiast was watching the Speed of Light train as it overtook the speeding bus. He also timed the train, but according to his stopwatch, the time for the train's leading edge to pass from the back of the bus to the front was about twice that recorded from the bus. Thus, it seems that the time difference between two events can vary for different observers, depending on their relative motion. Another way of thinking about it is that from the train enthusiast's viewpoint, the stopwatches on the bus were running at half the speed of his own.

The Average Lives of Muons

The particles that help prove the validity of special relativity are called muons. A stationary muon has an average lifetime of about two millionths of a second. But when moving at close to light speed (as they often do), muons can extend their average lifetime to ten millionths of a second or more.

Space-Time

Within a few years of Einstein publishing his ideas about relativity, a colleague of his, mathematician Hermann Minkowski, introduced the concept of "space-time." This was based on the idea of describing events in the Universe in four dimensions—three of space and one of time—instead of the more conventional three dimensions. The purpose of introducing space-time was to make it easier to measure the separation between events. Under the old measurement system, when two events occurred in separate places, both the distance and time between them were ambiguous because, as shown by the special theory of relativity, observers traveling at different speeds measure different distances and time intervals. Using a space-time system, the separation between any two events can be unambiguously described by what is termed a "space-time interval."

Proof of Relativity

Einstein published his special theory of relativity in 1905 and his general theory in 1915, but it was years before the theories were proved to be correct. One of general relativity's predictions—that massive objects such as stars can bend light rays passing near them—was proved in 1919, through observations of light rays from distant stars passing close by the Sun. Astronomers had to wait for a solar eclipse to make these observations. Special relativity's prediction that time passes more slowly for objects moving close to the speed of light was confirmed in 1940: certain unstable elementary particles were found to take longer to disintegrate, in other words to "live" for longer, if they were moving at close to light speed. Another of special relativity's findings, that mass and energy are equivalent, was proved (in a sense) by the detonation of the first atomic bomb in 1945.

General Relativity

The general theory of relativity is a more comprehensive theory than special relativity and deals with accelerated motion (where the velocity of objects changes). Importantly, general relativity is also a theory of gravity. It is based on the idea that the effects of being accelerated (for example, by a rocket) and the effects of being in a strong gravitational field (for example, from standing on the surface of a large planet) are indistinguishable. Starting from this principle, Einstein developed an argument that gravitational effects are caused by large concentrations of mass-energy causing local distortions in the shape of space-time. According to this idea, a planet in orbit around a star follows a curved path not because the star is pulling on the planet, but because the mass of the star has distorted space-time, and that the shortest path for the planet to take though this warped region is a curved one. General relativity also proposes that a ray of light is bent as it passes a large mass.

Distorting Space-Time

The central theme of Einstein's general theory of relativity is that massive objects produce distortions in space-time, and that gravity operates through these distortions. One way of visualizing space-time is as a flat rubber sheet in which massive objects make dents. In this analogy, light passing by a massive object has its straight-line path deflected as it follows the dent the object makes in the space-time sheet.

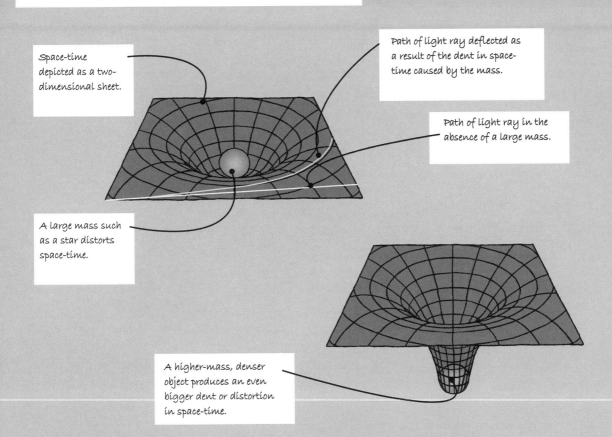

Space-time depicted as a two-dimensional sheet.

Path of light ray deflected as a result of the dent in space-time caused by the mass.

Path of light ray in the absence of a large mass.

A large mass such as a star distorts space-time.

A higher-mass, denser object produces an even bigger dent or distortion in space-time.

Although these concepts of general relativity might seem far-fetched, the application of the math associated with them works perfectly. For example, because it orbits so close to the Sun, the planet Mercury always moves in a very strong gravitational field (or according to general relativity, strongly curved space-time). Its orbit displays oddities that could never be accounted for by classical mechanics but are explained perfectly by general relativity. The fact that rays of light are bent as they pass close to stars and galaxies has been proved beyond doubt and occurs exactly as predicted by general relativity. General relativity also provides descriptive frameworks for gravitationally extreme phenomena, such as black holes, and provides models for the structure, development, and eventual fate of the Universe.

Before Einstein proposed his general theory of relativity, space and time were regarded as a sort of arena in which events take place. Scientists now realize that they are dynamic flexible entities that are influenced by energy, forces, and mass.

• Quantum theory is an attempt to describe the interaction of matter and energy at very small, subatomic levels.

• An important aspect of quantum theory is that forms of energy usually regarded as waves sometimes behave like particles, and particles sometimes behave like waves of energy.

• In the quantum world there are few impossibilities, but only events with very low probabilities, and these events occur regularly and predictably.

• One interpretation of quantum theory states that if an object can be in different states, then for as long as no-one looks, it is in both states simultaneously.

Quantum Theory

Quantum theory, or quantum physics, is an extremely important area of modern physics concerned with the behavior of matter and energy at tiny scales—atom-sized or smaller. Like relativity (see pages 164–167), quantum theory takes over from classical physics in certain situations—in this case in the minuscule world of subatomic particles, sometimes called the quantum world.

Quanta, Waves, and Particles

One of the fundamental ideas of quantum theory is that energy is absorbed, emitted, or transferred between matter in tiny, separate packets, called *quanta*, rather than continuously. Quanta of light and other electromagnetic radiation are called photons and can be considered to be particles. Light and other forms of electromagnetic radiation therefore have particle-like properties, although they also have proven wave-like properties. Conversely, subatomic particles such as electrons have demonstrable wave-like properties as well as their particle-like properties. This double nature of energy and particles in the quantum world is referred to as wave–particle duality.

Schrödinger's Cat

There is a clear but puzzling separation between the quantum world and the larger-scale, or macroscopic, world we are all familiar with. Take, for example, the thought experiment known as Schrödinger's Cat. This was devised to show that effects applying in the quantum world lead to absurdities when applied to the macroscopic world. In this hypothetical experiment, a cat is placed in a sealed box together with an apparatus for measuring whether an event with a 50:50 chance of happening—the decay of a radioactive isotope—actually occurs. If the event happens, a poisonous gas is released in the box and kills the cat. If it does not, the cat survives. Until the box is opened, it is not known whether the isotope decayed or not. Thus, the fate of the cat is also not known. According to quantum theory, this means that while the box is still unopened, the cat is in a mixed state—50 percent alive and 50 percent dead. But since nobody believes this is a valid outcome, the question then arises as to the scale at which quantum theory breaks down and "classical" physics takes over.

The Fuzzy Nature of Subatomic Particles

Consistent with their wave-like character, particles such as electrons have a vague nature in terms of their position in space, their shape and size, and their momentum. In particular, it is not possible to know about all the properties of an electron, such as its position and momentum, at the same time (this is called the "uncertainty principle"). In fact, a particle such as an electron may be best thought of as a fuzzy cloud, called a probability cloud, the location of the particle at any spot in the cloud being expressible only as a probability. Because of their fuzzy nature, subatomic particles can do some surprising things such as quantum tunneling (see opposite). This is of particular importance to electronics and computer science.

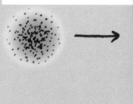

THOUGHT EXPERIMENT:
A "what if" sort of experiment, performed in the mind rather than in a laboratory.

MOMENTUM: The product of an object's mass and its velocity.

Jargon
buster

Quantum Tunneling

In the quantum world of objects the size of atoms or smaller, things can happen that just aren't possible in the large-scale, macroscopic world with which we are familiar. One example of this is the phenomenon called quantum tunneling.

In the quantum world, an electron is fired toward a barrier that repels the electron. An electron has a much vaguer shape than a ball—in fact it can only be represented by a fuzzy sphere. The electron could be anywhere in this sphere, the probability of it being at any one spot varying with the density of the cloud at that spot.

In our large-scale world, suppose a ball is thrown or kicked against a barrier such as a wall.

The ball doesn't have enough energy to break through the wall, so it rebounds.

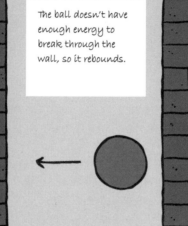

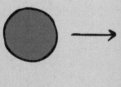

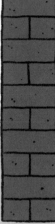

By the time the electron has reached the point where it is being repelled by the barrier, a small part—say, about one tenth—of its probability cloud has already passed through the barrier. This means that there is already about a one tenth chance that the electron has already passed through the barrier!

Because everything in the quantum world is decided on probabilities, what this means in practice is that about nine times out of ten the electron will be repelled by the barrier, but will pass, or "tunnel," through it about one in ten times. In other words, if 100 electrons are fired at the barrier, about 10 might pass through, even though each electron has insufficient energy to break through the barrier.

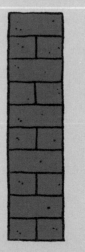

9/10 chance

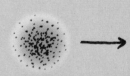

1/10 chance

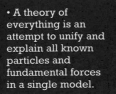

- A theory of everything is an attempt to unify and explain all known particles and fundamental forces in a single model.

- Numerous theories of everything have been proposed so far, but none has been proven by experiments.

- Physicists have identified particles that transmit three of the four fundamental forces of nature, but a transmitter for the force of gravity remains elusive.

- Currently some of the best contenders for a satisfactory theory of everything are string theories, which propose that subatomic particles are formed from minuscule vibrating strings.

Theories of Everything

A theory of everything is an overall model that explains all fundamental forces of nature and all the different types of particle that exist in the Universe. In particular, it is an attempt to unify gravity with the other three fundamental forces—the strong force and weak interaction (two forces that operate inside atoms), together with the electromagnetic force, which control phenomena such as electricity and magnetism, chemical structures, light, and much else.

Why is a "Theory of Everything" Needed?

Of the two major theories of physics developed during the 20th century and considered on the previous pages—relativity and quantum theory—the first (relativity) is good at explaining objects that are extremely massive or moving at ultra-high speeds, whereas quantum theory helps explains the subatomic world. The two theories might be expected to dovetail neatly together, but unfortunately they don't, which has caused physicists tremendous problems and provides a major block to further progress in understanding the physical nature of the world.

The Elusive Nature of Gravity

Quantum theory has helped scientists develop convincing models for how three of the four fundamental forces of nature operate. It has also helped them to begin to unravel how these three forces might once have been combined in a single force that operates only at extremely high energy levels, as existed in the very early Universe. However, quantum theory has so far failed to explain gravity. Whereas physicists have found "messenger" particles that transmit the other three forces, they have been unable to confirm the existence of a gravity particle, or "graviton." Any theory of everything would need to find one and discover other common ground between gravity and the other fundamental forces.

GRAVITON: A hypothetical elementary particle that transmits gravity. The properties that a graviton would have are well understood (for example, it would have no mass), but despite a search lasting several decades, no graviton has so far been identified.

String Theories

Currently among the best contenders for a theory of everything are so-called string theories. These propose that subatomic particles consist of minuscule filaments called strings, which can be either open-ended or closed (tied at the ends) forming loops or rings. The different theories propose that these strings vibrate, and that the different types (modes) of vibration create the various particle types. For example, one vibrational mode might make a string act as an electron, another as a quark. The dynamics of strings can also be used to model all the four fundamental forces of nature—the properties of some closed strings can be used to explain the workings of gravity, and open strings can be used to model the other fundamental forces. Open and closed strings can split and combine with each other, and open strings can turn into closed strings, representing various types of particle interaction.

After more than 300 years, scientists know almost everything about what gravity does, but hardly anything about how its force is transmitted. Theories of everything, such as string theory, are an attempt to remedy this situation.

How Big Are the Strings in String Theory?

Strings are almost too small to imagine. We have already seen (on page 13) that if an atom were the size of a sports stadium, then its nucleus would be roughly the size of the ball in the referee's whistle. Strings are much smaller than that. In fact, a sports stadium is too small a structure to be used in this comparison. Something much larger is needed. If an atom were, say, the size of a small town, would a string be the size of a pinhead? No. The string needs to be compared with something even bigger. If the atom were as big as a country? As big as planet Earth? Could we say the string would then be the size of an "s" on this page? No. Still nowhere near. To get anywhere near an idea of the size of a string, the atom would need to be so big it would be bigger than our own Solar System. Its diameter would take it all the way to the next star, which is more than 250,000 times farther away than the Sun. And even then, the vibrating string inside it would still be much smaller than the width of a human hair. That is, as they say, hard stuff to get your head around, and to many the whole idea may seem ludicrously fanciful. But many physicists think this model may be the one that works. The math seems to stand up, and the theory goes some way to explaining how all four of the fundamental forces work—including the most elusive force of all, gravity.

Index

Credits

Author's Acknowledgments

I would like to thank Steve Parker for his significant contribution and assistance with Chapter Seven, on health matters, Cathy Meeus for her editorial input, and Michael Chester for his illustrations.

I would also like to thank various people at Quarto Publishing: Kate Kirby, Paul Carslake—particularly for his input on illustration ideas—Caroline Guest and Moira Clinch, and last but not least Chloe Todd Fordham for her encouragement and considerable forbearance.